W9-BUJ-892

Chicago
Jazz

Chicago

Jazz

A
Cultural
History
1904–1930

WILLIAM HOWLAND KENNEY

New York   *Oxford*

OXFORD UNIVERSITY PRESS

1993

Oxford University Press

Oxford   New York   Toronto
Delhi   Bombay   Calcutta   Madras   Karachi
Kuala Lumpur   Singapore   Hong Kong   Tokyo
Nairobi   Dar es Salaam   Cape Town
Melbourne   Auckland   Madrid

and associated companies in
Berlin   Ibadan

Library of Congress Cataloging-in-Publication Data
Kenney, William Howland.
Chicago jazz : a cultural history, 1904–1930
/ William Howland Kenney.
p. cm. Includes bibliographical references and index.
ISBN 0-19-506453-4
1. Jazz—Illinois—Chicago—History and criticism.
2. Chicago (Ill.)—Social conditions. 3. Chicago (Ill.)—Race relations.
4. Chicago (Ill.)—Popular culture. I.Title.
ML3508.8.C5K46   1993
781.65'0973'1109042—dc20   92-27397

1 3 5 7 9 8 6 4 2

Printed in the United States of America
on acid-free paper

For Françoise and Mélanie

Acknowledgments

MY INTEREST IN the cultural history of jazz was sparked by historian Neil Leonard many years ago at the University of Pennsylvania. In a small way at least, I hope this volume will indicate his enduring inspiration. The work of Lewis A. Erenberg deeply influenced my initial thoughts about a model for this study. Two Kent State University colleagues—Professors Doris Y. Kadish and August Meier—played important roles by taking precious time from their own research to help me to improve the focus and organization of my chapter drafts. Their intellectual discipline and generosity have acted as a constant source of scholarly inspiration, while their faith in this project's potential, like that of editor Sheldon Meyer, has been vastly encouraging.

I gratefully acknowledge fellowship support for this book from the American Council of Learned Societies and the National Endowment for the Humanities. Both awarded summer Grants-in-Aid in 1989 for the collection of source materials in Chicago. The Research Council of Kent State University provided an Academic Year Research and Creative Activity Appointment in 1989 and travel funds to return to the archives thereafter.

The staffs of several libraries and archives, worked with remarkable skill to make my job easier. These organizations are listed on page 181. I would particularly like to acknowledge, however, the permission granted by Ron Greele of the Columbia Oral History Collection and Mary Ann Bamberger of the University Library, University of Illinois at Chicago to quote from the collections under their management. Michael Cole, Linda Burroughs, and Mary Taylor of the Periodical Information and Access Department at Kent State did a skillful job of making available to me crucial interlibrary loan, periodical, and microfilm documents.

I would also like to express my indebtedness to the dedicated jazz record collectors who went out of their ways to tape Chicago jazz records for my use. Joel O'Sickey again provided invaluable assistance in this regard, as he has on past projects, and I'm particularly grateful to him for sharing his considerable knowledge of the phonograph records with me. John and Nina Steiner offered me their hospitality while I learned from the Steiner Collection. Jim Stincic, John Richmond, Larry Booty, and John Bitter all provided records and tapes of Chicago jazz. Jazz guitarist Brad Bolton diagrammed some of the more complex big band recordings. All these individuals proved the strength of the jazz community and the role it can play in helping to recreate a cultural history of jazz. Above all, Françoise Massardier-Kenney, who has shone a bright and encouraging light on my path to and through this project, has my heartfelt thanks.

Kent, Ohio   W. H. K.
June 1992

Contents

Introduction

THIS BOOK ADVANCES an interpretation of the cultural history of jazz in Chicago from its origins to the Depression. In recent decades musicologists and music critics, whose major findings will be incorporated into the present study, have demonstrated the musical depth and artistic originality of jazz, tracing the lines of influence running from one outstanding jazz artist to another.[1] But historians have become increasingly active in the scholarly analysis of jazz. Cultural historians, in particular, have traced the relationship of jazz to changing African-American sensibilities as they moved from slavery to freedom, as well as studying the interactions of jazz and race within the cultural patterns of the dominant white society at given moments in American history.[2]

A closer analysis of one particular time and place, widely recognized to have been important to the evolution of jazz, will help to re-create a clearer sense of the particular social and cultural circumstances—what anthropologist Clifford Geertz called the "flow of social action" and the "conceptual structures"—within which jazz musicians molded their music. Leroy Ostransky has demonstrated that several powerful historical forces, which were at work early in this century in major American cities, created the political, economic, and racial preconditions for the development of jazz.³ This study tries to reconstruct some of the interactive process between music and the specific night-life institutions that presented jazz in twenties Chicago. The job, as South Side newspaper columnist and orchestra leader Dave Peyton insisted, created bands and held them together. Cabaret, dance hall, and vaudeville theater employment gave life to jazz groups:

> The job makes the orchestra. If you lose the job and loaf a few
> weeks, you haven't any band. Our field is a narrow one. Your men
> can't afford to loaf long and the first bidder takes them away from
> you. The job is what you want to worship.⁴

The distinction often drawn between the music that jazz musicians had to play on the job and the music that they would personally have preferred to play should not be allowed to obscure the depth of jazz's involvement with its immediate surroundings. Regular employment in Chicago's cabarets helped to shape the evolution of instrumental techniques, solo flamboyance, skills in accompaniment, jazz band repertoires, and new concepts in composing and arranging.

The field of musical expressiveness suggested by the word "jazz" presents certain complications not usually encountered in cultural history. The term's exact meaning has been notoriously elusive, covering, both then and now, a wide variety of different musical styles and provoking substantial disagreement. I have retained much of the definition of jazz offered by Joachim E. Berendt, who focused on the artistic confrontation of blacks with European music within the United States. Berendt emphasized a series of musical characteristics—rhythmic swing, creative spontaneity, improvisation, and individualistic instrumental sonorities—which characterized African-American jazz music.

The notion of jazz as a musical art form is central to the field, but should be broadened to include white artistic confrontations with African-American musical traditions. Based on facts found in the primary sources, this book affirms that black South Side Chicagoans were chronologically well in advance of the city's white musicians in developing the music, but whites soon made their contributions, too. Members of all races, however, felt many of the impulses that generated jazz. The primacy of black jazz in Chicago during the twenties resulted from craftsmanship developed in response to musical ambitions and economic opportunities in a white society that had long expected and encouraged African-Americans to make music. As a 1906 newspaper article from the white Chicago *Tribune,* reprinted in the black Chicago *Broad Ax,* put the matter:

> ... the Negro has a future in music .... there is no prejudice
> against the Negro in music ... He need not fear that race prejudice
> will antagonize him. Music is the universal art and language and
> begins where speech ends.[5]

Chicago Jazz is on one level (some would say the only important one) the story of the musical creativity of a small group of South Side Chicagoans who pioneered jazz's major musical breakthroughs, but the word also carried, in the 1920s, a much broader, cultural signification. Most Chicagoans were only marginally aware of the complex and subtle details of African-American encounters with European music. While it is true that Chicago's great jazz musicians of the 1920s often succeeded in recording music with an appeal that lasted well beyond the moment of its creation, it is also true that those same artists habitually performed as musical entertainers when creating their jazz art. As a result, during the late nineteen teens and twenties, Chicago jazz nearly always stimulated and responded to Jazz Age cultural sensibilities in Chicago and the nation. Therefore, a distinction can be made between "jazz," as innovation in musical art, and the phrase "jazz age," that can be used to describe "Roaring Twenties" social dance music and associated activities, such as going out to dance halls and cabarets, going to the movies, dressing like "sheiks," "shebas," and "flappers," and drinking bootleg gin.

These activities, like jazz itself, were "jazzy" urban behaviors that expressed the excitement, adventure, glamour, sensuality, and daring stimulated in young urban Americans. Jazz Age sensibilities did not require any overt confrontation at the intersection of race and music, although those closest to Chicago jazz in the twenties insisted that they discovered their most memorable musical experiences there. Chicago Jazz proved capable of expressing a range of emotions that stimulated and reflected the excitement of the Roaring Twenties and sometimes achieved a level of musical expressiveness which attracted critical praise long thereafter. Some musicians and groups shaped these musical expressions more distinctively than others; some played more consistently in clubs associated with jazz, while others played some jazz numbers among other styles of popular music in venues not exclusively identified with jazz.

All of the musicians and audiences of that time were enmeshed in various mixes of the popular emotions, feelings, and tastes which characterized the Roaring Twenties. Jazz in Chicago gave particularly sharp and memorable musical expression to feelings of giddy excitement and rebellious daring, stimulating powerful emotional experiences which permanently shaped our collective memory of that time and place. Although individual musicians and groups placed themselves differently within the sensibilities of the time, they all were inextricably involved with post-World War I economic prosperity, a heightened interracial awareness, the excitement of city life, urban machine politics, and the ongoing rebellion against Prohibition moralism.[6]

In the pages that follow, I have tried to describe the ways in which culture, race, and music acted on one another to draw upon specific elements of African-American musical traditions in order to create jazz. Traditional white perceptions of racial characteristics and the craftsmanship of twenties jazz musicians combined to create exotic night-time worlds of urban excitement; the music itself simultaneously stimulated emotionally charged moments of social daring and sensations of racial reconciliation through ritually ordered rhythm and harmony. Developed as a way to make oneself heard within the city, South Side Chicago jazz, an object lesson in economic and cultural adaptation, served for many who heard it to affirm that a pluralistic urban society was possible.

Chicago has an especially broad range of primary and secondary sources dealing with the historical jazz scene of the 1920s. Newspapers, trade papers, phonograph records, transcribed oral interviews, and autobiographies provide the means to discover how music, race, and culture interacted in that urban setting. In addition, the history of early twentieth-century Chicago has produced an unusual number of studies of major social developments—urban vice, racial relations, dance halls, and spreading slums—that were acknowledged at the time to have had an association with jazz. In a sense, therefore, the sort of cultural history of Chicago jazz which I have tried to write has been waiting to be written.

This examination of jazz during a major period of its gestation seems in one sense to confirm that the music itself emerged from a blending of "lowbrow" cabaret culture and elements of "highbrow" musical tradition.[7] At least from the Chicago experience of its early innovators, jazz provides a remarkable example of aesthetic activity in which "high" and "low" cultures are not necessarily antagonists. For instance, jazz musicians readily borrowed from a variety of sources in refashioning the music they had brought to Chicago. Jazz in 1920s Chicago was neither just folk music, nor merely commercial entertainment, nor solely concert hall art, but instead it was a synthesis of these elements. In the process, it created an original, indigenous, richly varied music form. The stock market crash, the repeal of Prohibition, and the Depression of the 1930s were factors that helped destroy this Chicago jazz tradition and in the process remove the recordings of the period from their original cultural context.

This study also demonstrates that beneath the heated debates over jazz in 1920s newspapers, art magazines, and education journals— debates that treated jazz as a "revolutionary" social and cultural phenomenon—there were groups of people who did not normally write about jazz, but just played, promoted, and listened to it in clubs and dance halls around the city. Whatever musical borrowing jazz made from "highbrow" sources, and in spite of its popularity with slumming members of the upper middle class, Chicago jazz of the twenties was firmly rooted in a tough inner city world of working-class entertainers.

F. Scott Fitzgerald, who first applied the term "Jazz Age" to the 1920s, traced the unfolding of disturbing new sensibilities of the twenties within the social worlds of college campuses, country clubs, and elegant hotels. Fitzgerald wrote little at all about jazz music, its musicians, or the important jazz clubs. I have tried, in this book, to recapture some of the substance and flavor of the more immediate inner-city world of jazz cabarets and dance halls in twenties Chicago. Jazz looks different, somewhat less "revolutionary" and more evolutionary from this other perspective, one closer than that of twenties social critics and writers to the worlds of the musicians themselves, who actually created jazz in the exciting, enterprising world of twenties Chicago.[8]

Chicago
Jazz

1

South Side Jazz: Cultural Context

DURING THE 1920s, Chicago provided a major focal point of innovative jazz performance in America, drawing to itself musicians from throughout the Midwest and the Deep South. Jazz legend has focused on the U.S. Navy Department's November 12, 1917, closing of "Storyville," New Orleans' celebrated red light district, as the fundamental impetus to the emigration of jazz musicians from the Crescent City. Since some of these subsequently famous New Orleans musicians also performed on the Streckfus Line's Mississippi River paddle wheelers, jazz also has been said to have "come up the river from New Orleans." But the closing of the official vice districts in cities like New Orleans merely dispersed their activities into the surrounding neighborhoods. Both bordello and dance hall continued to flourish, as the official closing of Chicago's "Levee" district in 1912 amply proved. Moreover, as Richard Wang has pointed out, jazz legend ignores North American geography: the Mississippi River doesn't flow through or even very near Chicago.[1]

Jazz musicians were not so much pushed from their many different homes throughout the country as pulled to Chicago. Popular musical development actually began there well before World War I, stimulating and reflecting a mounting popular "craze" for social dancing. During the twenties, what Marshall Stearns called a "peak of jazz intensity" developed in New Orleans, Chicago, New York, and Kansas City. Chicagoans listened and/or danced to jazz in a variety of settings: white, black, and black-and-tan cabarets, roadhouses, vaudeville and movie theaters, dance halls, excursion boats and trains, and private parties. Journalists for the entertainment trade papers underscored Chicago's leadership in cabaret orchestras; they marveled that jazz and dance bands had become a dominant force in vaudeville, replacing vocal acts as the major medium for creating hit tunes.[2]

Chicago's booming market for entertainment encouraged intense musical competition and led to key musical discoveries that propelled jazz forward during the 1920s as a new American art form. Most of the city's creative musical advances emerged from African-Americans, who were crowded into a racial ghetto on the South Side; black musicians created many musical innovations in which white Chicago musicians and social dancers found their inspiration. Some of these daring creations were captured on influential phonograph records. Louis Armstrong's Hot Fives and Hot Sevens, Jelly Roll Morton's Red Hot Peppers, and critically acclaimed recordings by Joseph "King" Oliver, Jimmy Noone, Johnny Dodds, and Earl Hines provided enduring evidence of Chicago's importance in jazz history by defining major directions of jazz development for years to come.

The jazz music played by South Side Chicagoans sprang from a self-conscious evolution in African-American musical and show business traditions. Born of a combination of folk blues, marching band tunes, social dance music, popular songs, and ragtime, jazz found expression in twentieth-century urban adaptations of nineteenth-century minstrel and vaudeville forms. Like ragtime, jazz expressed two important facets of African-American sensibilities: first, such music was a valued element of black cultural life; and, second, certain kinds of musical activity were recognized by the surrounding white society as appropriate for blacks. The two dimensions combined to create strong foundations for making a creative popular music responsive to city life.

Jazz on Chicago's South Side was deeply woven into a fabric of economic and political activities designed to improve the standard of living and political power of the black community. South Side jazz, for example, was closely allied with a dynamic pattern of entertainment enterprise which, given major racial constraints on black economic activities, played an even more important role in the South Side community than might otherwise have been the case. Barred from most professional schools and corporations by reasons of color, entrepreneurs on the South Side focused an unusual amount of creative energy on such entertainment enterprises as cafes and saloons, pool halls, gambling, bootlegging, vaudeville, popular music making, and such fast-developing enterprises as the production of phonograph records and movies, and professional boxing, baseball, and football. Jazz in twenties Chicago closely associated with these kinds of activities.

At the same time, these enterprises intersected with the gathering political ambitions of the black community in Chicago. The rise of the South Side cabarets and jazz paralleled the determined and successful effort to elect the first African-Americans as City Council Aldermen. Chicago's Second Ward, where most of the leading cabarets were located, probably contained a majority of black voters by 1915 when, amid heated newspaper commentary, Oscar DePriest was elected Chicago's first black Councilman. In 1918, another African-American—Louis B. Anderson— was elected to replace DePriest, while Major Robert R. Jackson won the second of the two City Council seats allotted to the Second Ward. The leading clubs in which the famous black ragtime and jazz musicians played were owned and/or managed by black Republican party organizers, who used the musicians, their music, and popular musical entertainers to attract and to focus the attention of potential black voters. Jazz musicians, often thought to be apolitical, nonetheless helped to build, on the South Side at least, pride in the economic and political advances of the race. They also contributed their music to the ceremonial political meetings held on the South Side between Chicago's black and white politicians.[3]

The institutional roots of Chicago jazz reached back to a turn-of-the-century night club, gambling hall, theater, and political hot spot called the Pekin Inn, at 2700 South State Street. It was the most important South Side Chicago club and musical theater before 1910 and the first to employ musicians who were closely associated with ragtime and pre-jazz popular music. The Pekin began as a popular beer garden serving blacks and whites, but on June 18, 1904, was revamped as the Pekin Inn, "a cabaret better known as a Music Hall." On either March 17 or 30, 1906 (depending upon which source one reads) the Pekin was reopened as a 1,200-seat "theater for Colored people of this city," owned by Robert T. Motts, who had been born on June 24, 1861, in Washington, Iowa, and had come to Chicago in 1881. Joining forces with Samuel R. Snowden and William Beasley, Motts had made 480 South State Street a center for the sporting set before taking over the Pekin Inn, and had invested his gambling profits in South Side real estate. At the same time, Motts, who became an important South Side politician, used his club to mobilize the black vote. He paid some of his customers $5 per day to help the Second Ward Aldermen register black voters and see that they voted on election days.[4]

A black entrepreneur like Motts had to devise strategies for balancing economic with racial interests: whites generally had more money to spend than blacks, and therefore could easily account for the financial success of such an enterprise; but blacks, with less discretionary income, nonetheless dominated the Pekin's neighborhood creating contrary pressures for racially responsive entertainment policies. An influential pre-World War I report on the South Side black community indicated that black business succeeded financially only when two-thirds of the customers were white. A close observer of South Side cabarets insisted, "there is no reason to draw any color line when colored people have to struggle to make a living." The Pekin Inn and places like it therefore traditionally served the "sporting fraternity," an informal brotherhood of pleasure-seeking bachelors of both races. The sporting set included slumming young upper-class whites, who lined up at the bar and gathered around the gambling tables with downtown politicians, artisans, actors, and immigrants to the city from many lands.[5]

Robert Motts seems to have succeeded remarkably in appealing to race pride. The Chicago *Broad Ax,* an old-fashioned newspaper that opposed most of the social trends of the jazz age, compared him to Benjamin Banneker, Toussaint L'Ouverture, and Booker T. Washington and hailed him as "the new Moses of the Negro race in the theater world." Motts's Pekin was touted as "the finest and the largest playhouse conducted by Afro-Americans in the United States." The first show produced there in 1906 was called "The Man from Bam," a musical comedy with a book by Collin Davis, music, which included "The Rag Time Ballet," by Joe Jordan, and lyrics by Arthur Gillespie. Soon after its opening, Motts donated the premises and "the famous Pekin Orchestra" for a Grand Benefit for the Frederick Douglass Center. Jane Addams of Hull House, who sponsored a special program by Soper's School of Dramatic Art, and Mrs. Ida B. Wells-Barnett, who defended the Pekin Theater against criticisms from local ministers, placed their stamp of approval on Motts' venture. As long as Robert Motts lived, the Pekin

Theater was a self-consciously black Chicago entertainment institution that employed African-Americans in all capacities and charted an entertainment strategy which responded to the ambitions of African-American entertainers.[6]

The tension between economic necessity and racial solidarity was ongoing, though. In April 1910 the Chicago *Defender* wrote that Motts afforded blacks the possibility of enjoying nationally recognized vaudeville acts of both races "without going downtown . . . ," but the newspaper added that "[he] caters much at present to white people" and urged that he take care to pay his black acts as much as his white ones. In the months that passed thereafter, Motts was reported to have traveled to inspect the Howard Theater in Washington, D.C., where what the newspaper called a "Hebrew" proprietor was offering black audiences "[racially] 'mixed' vaudeville." Motts subsequently steered the Pekin more successfully along the tricky inter-racial path, away from a complete reliance on white acts and audiences and toward greater encouragement of race-oriented productions and the professional ambitions of black actors and entertainers. He presented legitimate plays like "Tallaloo" and "Carib," with African-American themes and produced and acted by the Pekin Stock Company of local black actors, entertainers, and musicians.[7]

Robert T. Motts died in 1911. At his funeral, he was eulogized by the Rev. W. D. Cook and the Rev. H. J. Cullis, with leading South Side banker Jesse Binga as an honorary pall bearer. Until World War I the Pekin was owned by Robert's sister Lucy Motts, who married Dan Jackson, the proprietor of the Emanuel Jackson undertaking business and later a leading South Side politician; thereafter a series of whites, among them George Holt, Frank Haight, Thomas Chamales, William Adams, and Walter K. Tyler, subsequently owned the Pekin Inn. According to folklore that circulated among those in the inexpensive balcony seats, "the Ghost of Robert Motts" caused the theater's curtain to stick shut on each opening night under new white management, since Motts was said to have requested on his deathbed that the Pekin remain always under black ownership. The theater seats were removed in the summer of 1916 and replaced by a dance floor, and the cabaret finally became a gangland hangout, where two policemen were shot in 1920. The building was eventually sold to the city and became a police station.[8]

The musicians who accompanied entertainers at the Pekin Inn and Theater and who performed their own musical acts there shared the

institution's orientation toward black entertainment and black audiences.
Among the Pekin Inn's most influential early musicians, composers as
well as instrumentalists, were Joe Jordan, who served as musical director
of the vaudeville and legitimate theater productions from 1906 and Tony
Jackson, who wrote the famous song *Pretty Baby.* Jackson apparently
arrived in Chicago around 1912. He also appeared at the Congress Hotel
and Russell and Dago's Elmwood Cafe, while Jordan appeared at clubs
like the Pompeii Buffet and Cafe and the Deluxe Cafe in the waning
years of the ragtime era just before World War I.[9]

People of both races also came to the South Side to hear clarinetist
Wilbur Sweatman, a vaudeville/novelty instrumentalist who presented
mixed programs of classical music, gypsy airs, and hot syncopated numbers to Chicagoans as early as 1906. Sweatman was famous for playing
"The Rosary" on three clarinets simultaneously and in harmony. He
performed with William Dorsey at the piano and George Reeves on
drums at the Pekin Inn, the Monogram Theater, and the Grand Theater,
the South Side's largest vaudeville/movie theater. Sweatman's clarinet
style was so unprecedented in Chicago that even musically sophisticated
black listeners found it strange:

> He was such a hit with his queer style of playing "Hot Clarinet"
> that Broadway [subsequently] went wild about him. People of both
> races came to hear this three piece orchestra play jazz music, al-
> though they didn't call it jazz then. They called it "hot music."
> Sweatman produced the weird, eerie tones on the clarinet that sent
> thrills through the listener. He was a sensational, rapid, clever ma-
> nipulator of the clarinet.

Sweatman later claimed to have recorded the first jazz records ever made
(in 1912 for Columbia Records), and advertised himself as "Originator
and Much Imitated Ragtime and Jazz Clarionetist...."[10]

With the death of Robert Motts in 1911, leadership of the South
Side's developing cabaret culture, as well as of its political ambitions,
passed briefly to black boxing champion Jack Johnson, who had used his
immense popularity as early as 1910 to campaign for Edward H. Wright,
the first black candidate for City Council. After defeating boxer Jim
Flynn in Las Vegas on July 4, 1912, Johnson opened the Cafe de Champion at 41 West 31st Street in "the Old Palace [theater]" between Armour
Avenue and Dearborn Street. A pianist was featured at the grand piano

in the upstairs cafe, and at the club's gala opening, W. H. Taylor's Orchestra presented an "elaborate program of music and song in the downstairs buffet." Johnson's influence on the developing patterns of musical entertainment on the South Side was soon limited, however, by his indictment on November 7, 1912, by a federal grand jury on a charge of trafficking in women. The champion, characterized by the *Defender* as a martyr to white racist resentments of his defeat of the white boxer Jim Jeffries, fled to Europe, although his club remained open.[11]

During this same period, Thomas McCain's Pompeii buffet and cafe at 20–22 East 31st Street, at the 31st Street elevated station, and Dago and Russell's Elmwood Cafe presented such leading musical entertainers as Tony Jackson, Ferd "Jelly Roll" Morton, drummer Manzie Campbell, and the highly regarded tenor vocalist and durmmer Ollie Powers. The Pompeii, which was renamed the Richelieu in 1914 when Jelly Roll Morton became its musical director, was known as a "hang-out for theatrical people"; the Elmwood attracted attention for its late Sunday afternoon concerts and Tuesday matinees, both managed by Ollie Powers. Even when owned by whites, clubs like these encouraged black musicians and entertainers who were "... closed-out of even middle-level vaudeville and theater work."[12]

By 1915, leadership of the South Side cabaret scene passed to Henry "Teenan" Jones, who had arrived in Chicago from Watseka, Illinois, in 1876. Jones had run the Senate Buffet and the Lakeside Club in Hyde Park for sixteen years before being forced out when that area became an all-white residential neighborhood. He then moved further north, joining with Art Codozoe and J. H. "Lovie Joe" Whitson to own and run the Elite Cafe at 3030 South State Street, next door to the "old" Monogram Theater, a major center for down-to-earth black vaudeville and musical entertainment that subsequently moved further south to 34th and State and became the "new" Monogram Theater. Jones, like Robert Motts, was deeply involved with the South Side's dynamic combination of politics, gambling, and musical entertainment. A bar and restaurant, the Elite No. 1 also functioned as a small music hall; a space for dancing was enclosed by brass railings, which also served to divide performers from the customers at show time. Three to four hundred patrons could be served at a time and, as the newspaper put it: "The entertainers and the orchestra always hit it up pretty lively during evening hours."

Teenan Jones, who was described as a "bosom friend of the late Robert Motts, was president of "the Robert T. Motts Memorial Association" and the "Colored Men's Retail Liquor Dealers' Protective Association," a founder of the Great Lakes Lodge No. 43, L. P. B. O. Elks, and an organizer in the Republican party.[13] In 1915, he struck off on his own, opening Teenan Jones' Place, also known as "Elite No. 2," at 3445 South State Street, identified by its white tiled façade, and advertised as "the most elaborate emporium on the Stroll. Fine wines, liquors, and cigars; cafe and cabaret in connection." The music at Teenan Jones place, at the time of its opening at least, encompassed both classical and popular styles, making listeners feel, according to the Indianapolis *Freeman*, "like you were listening to a grand opera—the next minute to high class vaudeville." Like the Elite No. 1, which had been next door to the old Monogram Theater, Jones' place was nestled next door to the new Monogram Theater and became known as an actors' hangout.[14]

Teenan Jones was looked upon as a political power on the South Side at least until 1917. Then he was indicted for conspiracy, along with Alderman DePriest, and turned state's witness and confessed that he had been the head of a gambling syndicate. Jones admitted making monetary contributions to DePriest and Police Captain Stephen K. Healy, contributions that were construed to be protection money to keep the police away from his gambling houses. While confessing to the charge in return for immunity from prosecution, however, Jones still insisted that he had only been involved in making political campaign contributions. His contention was supported by DePriest's defense attorney Clarence Darrow and by the variety of Jones' other political activities on the South Side. DePriest was ultimately acquitted.[15]

At about the same time that Jones opened his Elite No. 2, Frank Preer and William Bottoms, with the assistance of Virgil Williams, opened the Deluxe Cafe at 3503 South State Street. Sometimes mentioned in the press as the LeLuxe, this cabaret featured vocalists Lucille Hegamin and Ollie Powers and strove to establish a reputation as a morally upright establishment where fighting was prohibited. When, just over one year later, Frank Preer suffered what one newspaper called "a nervous breakdown from overwork," Bottoms took over the club.[16]

Given the constraints on black economic and cultural initiatives in Chicago, clubs like the Pekin, Cafe de Champion, the two Elites, and the Deluxe played important roles in the ghetto's institutional life. They hired significant numbers of neighborhood men and women as floor managers, entertainers, musicians, cooks, barmen, waiters, doormen, and janitors, enough to contribute to the economic foundations of the South

Side. South Side cabarets also brought elements of urbane popular culture to the black ghetto. The Chicago *Defender,* which encouraged the popular culture of the twenties as much as the *Broad Ax* discouraged it, declared, with the sort of hyperbole reserved for race initiatives, that the Elite No. 2 "... [is] the most elaborate and well-appointed cafe and buffet in the country." Thanks to Jones, "... the race will have a Rector's or a Vogelsang's [elegant (white) lobster palaces] of its own ... a Mecca for High-Class Amusement."[17]

Jelly Roll Morton's early career in Chicago developed the most important of the South Side cabarets during 1914 and 1915. He played and/or organized vaudeville entertainment at the Richelieu, DeLuxe, and Elite #2 (mistakenly identified as the first Elite in the Lomax/Morton memoir). His Chicago activities enlivened the most influential of South Side Chicago's musical cabarets. His piano playing style best documents the transition from ragtime to jazz.[18]

The word "jass" first appeared in the city's black press in connection with the Pekin Inn, when, on September 30, 1916, the Chicago *Defender* used the word to describe music produced by black pianist-songwriter W. Benton Overstreet in support of vaudevillian Estella Harris at the Grand Theater. Harris, variously labeled as a "Coon Shouter" and "Rag Shouter," was now accompanied by a "Jass Band" which had been known as "the famous Pekin Trio" up to that time. They were a smash hit doing "Shima Sha Wabble," "Happy Shout," and "The New Dance"; one thousand fans were turned away at one late Sunday night show. Very soon thereafter, a variant spelling of the term—"Jaz"—was used in the Indianapolis *Freeman* to describe an instrumental group, John W. Wickliffe's Ginger Orchestra "Styled America's Greatest Jaz Combination"—which included Chicagoan Darnell Howard on violin and an "Entertainer or Interlocutor."[19]

The Great Migration of the years 1916 to 1919, which brought approximately 500,000 blacks from the southern states to northern cities (nearly one million more followed in the 1920s), greatly increased the demand for entertainment in these northern cities. The arrival in Chicago of over 65,000 blacks from Louisiana, Mississippi, Alabama, Arkansas, and Texas between 1910 and 1920 triggered Chicago's Jazz Age, for it expanded the city's market for racially oriented black musical entertainment and also intensified white Chicago's awareness of a growing black population. In the process, it created a broader market for black entertainment aimed at white audiences. During World War I, immigration from Europe slowed drastically, and many European immigrants moved back to their homelands. Northern industries like Chicago's Illinois Central Railroad, International Harvester, the steel mills, and the Swift and Armour stockyards and slaughtering houses actively recruited nonunionized black laborers to take their places. By 1920, over 100,000 African-Americans lived in Chicago, an increase of 148 percent in ten years.[20]

Most of the city's South Side jazz performers arrived between 1917 and 1921 at the height of this migration. Of the fifty-five black musicians, vocalists, and orchestra leaders closely associated with jazz in Chicago during the 1920s about whom information is available, nearly half arrived during or just after World War I. About the same percentage came from New Orleans. These migratory early jazz musicians moved about the country wherever work beckoned, often in company with the vocalists, mimes, acrobats, and buffoons of carnival, circus, minstrel show, and vaudeville troupes. Many of them had traveled extensively before settling into relatively lucrative, longer-lived Chicago jobs. The Original Creole Orchestra, usually cited as a major influence in the transition from instrumental ragtime to jazzband music, toured the nation on several vaudeville circuits for many years before two of its most influential musicians— the bassist/leader Bill Johnson and cornetist Freddie Keppard—settled in Chicago in 1918. Reed soloist Sidney Bechet, who arrived that same year, had played dances, shows, one-night stands, and dime stores all over Texas with pianist/composer Clarence Williams. At age fifteen, New Orleans trumpeter Lee Collins had worked for the Illinois Central Railroad in Cairo, and subsequently toured Alabama, Mississippi, and Forida before moving to Chicago.[21]

Immigration to Chicago thus did not necessarily signify uninterrupted residence; many jazz musicians discovered compelling economic and political reasons to keep a suitcase packed, and their music retained the subversive strains of social alienation. Some who played long, celebrated runs in Chicago traveled widely with vaudeville companies when club jobs grew scarce, as they did during the mild economic depression in 1921 and early 1922 and later when Prohibition politics closed certain cabarets; others worked extensively in St. Louis, Milwaukee, and other midwestern cities; the most famous worked for extended periods in New York, Los Angeles, and San Francisco, and a few even toured Europe and the Far East.[22]

Economic opportunity in a growing market for leisure time entertainment drew African-American musicians north from New Orleans into competition for the dollars that urban workers could devote to recreation in Chicago. From 1910 to 1916, New Orleans musicians had earned between $1.50 and $2.50 per engagement, plus tips, which could double that sum. After World War I, South Side Chicago cabarets paid sidemen a weekly salary of around $40 which also could be supplemented by tips. When groups of white Gold Coast socialites decided to "go slumming" in the cabarets, tips could amount to more than the regular pay. Orchestra leaders and solo stars, of course, earned more than sidemen.[23]

Like other immigrants, musicians felt the excitement Chicago could generate. Pianist Lil Hardin, who moved to Chicago from Memphis in 1918, recalled the northern city's magic many years later: "... I made it my business to go out for a daily stroll and look this 'heaven' over. Chicago meant just that to me—its beautiful brick and stone buildings, excitement, people moving swiftly, and things happening." Despite the bloody race riot of 1919, black immigrant musicians retained a guarded but persistent optimism born of job opportunities, higher wages, personal freedom, and the reassuring sparkle of cabaret lights and standing-room-only signs. They would have agreed with the grim optimism expressed by one anonymous prospective emigrant from the South: "I suppose the worst place there is better than the best place here."[24]

Many of those gripped by "the moving fever" on "The Flight Out of Egypt" arrived in Chicago as young adults who had left family ties and the older generation behind at least until they could get settled; they were eager to enjoy "the exhilarating feeling of liberation" from lives of southern dependency. Historian James Grossman has emphasized the "importance of bright lights and leisure opportunities ... to the migrants' image of Chicago as a freer environment than the rural or small town South." Temporarily freed from many of the traditional controls of family and church, unattached men and women between the ages of twenty and forty-four, who accounted for an unusually large proportion of the total black population, lived in the rapidly growing South Side rooming house district, where they created informal, transient relationships. In this shifting, restless world emerged the "rent party," a semipublic social occasion where young men and women paid a modest admission charge to dance to the latest phonograph records, a "boogie woogie" pianist or a small combo, drink inexpensive bootleg alcohol, and make friends. The jazz, like the alcohol, provided entertainment while helping the party's organizers to pay their rent.[25]

Workers with money to spend on entertainment also helped to create a market for more elaborate commercial ventures. During and after World War I, South Side Chicago, particularly "The Stroll," the bright-light district on South State Street, came alive with a fast moving, free-spending night life. In the early years of this century, South Side night life first focused on the intersections of South State with cross streets numbered in the twenties, the area of the old vice district called the Levee. This sin and entertainment district south of the Loop had grown up to service a transient population that circulated around the South Side terminals of the city's railroad, elevated, and trolley lines. The twenty-square block "Vice District" had comprised 500 saloons, 6 variety theaters, 1000 "concert halls," 15 gambling houses, 56 pool rooms, and 500 bordellos housing 3000 female workers. The showplace of the Levee had been the internationally famous Everleigh Club at 2131–33 Dearborn Street. Run by two Kentucky sisters, Ada and Minna Everleigh, the club had even used hidden wall devices to shoot perfume into the rooms. Thanks in part to the Everleighs' flaunting disdain of reformers, the Levee had been officially closed by Mayor Carter Harrison in 1912; but the move had succeeded only in moving such activities into other areas of the city and even into the suburbs.[26]

As the black population grew on the South Side, the center of the black bright-light district moved southward away from the area of the Pekin Inn on 27th Street to 31st Street (Elite Cafe No. 1, Grand Theater, Royal Gardens Cafe) and then to 35th Street, where much of the more influential jazz activity of the 1920s took place—in the Elite Cafe No. 2, DeLuxe Cafe, Dreamland Cafe, Sunset Cafe, Plantation Cafe, and the Apex Club.

White jazz personality Eddie Condon later claimed that in 1924–26, at the height of the jazz age, a trumpet held up in the night air of the Stroll would play itself. Stores remained open twenty-four hours a day to serve those enjoying urban life after years of rural tranquility. During the day, women wearing what the *Defender* called "head rags of gaudy hues" leaned from tenement windows while small groups of men asserted a more public presence on the sidewalks. At night the crowded sidewalks rang with music and laughter, the cabarets, vaudeville and movie theaters interspersed with "gaudy chile, chop suey, and ice cream parlors."

Thirty-fifth and State streets offered a cosmopolitan "Bohemia of the Colored Folks," where "lights sparkled, glasses tinkled," and crowds of people circulated, around the clock.

Langston Hughes, visiting from New York in 1918, recalled that "midnight was like day," even though electric streets lights were not installed by the city on State Street until late 1922! According to black dance band leader William Everett Samuels, some South Siders, working as postal clerks, mail order operatives, hotel porters, and a variety of other jobs during the day, often went home to bed at quitting time, arose at 2:00 a.m., dressed up in their finest clothes, hung out on the Stroll, sobered up in a steam bath, and returned to work.[27]

The Stroll and the southward-moving South State Street bright-light district, which would be centered at 47th and State by the end of the decade, created an enterprise in racial entertainment serving the tourist market on at least two levels: first, many black customers on the side-walks and in the commercial establishments were tourists from different points throughout the Midwest; secondly, the arrival of so many migrant blacks in Chicago stimulated the curiosity of whites, who came to the Stroll to experience something of "Race" life in the north.

To attract black tourists, Robert S. Abbott's Chicago *Defender* issued both a city and a national edition, the latter intended for broad circulation to readers throughout the Midwest and South. Both editions touted the Stroll as "a Mecca for Pleasure" and, likened South Side Chicago to Rome, Athens, and Jerusalem, a center of cultural attraction for African-Americans, where no one need fear "racial embarrassment." The paper reported that the Stroll was regularly "filled with tourists and travel-ers," particularly when the black delegations of such national organiza-tions as the Republican party and racial splinter groups like the African-American Elks met in Chicago. The Indianapolis *Freeman* regularly covered Chicago news in the manner of a Windy City newspaper, and its readers learned that the best and most sophisticated African-American cabarets and vaudeville theaters were in Chicago and New York.[28]

Racial tourism among whites was governed by white American atti-tudes toward African-Americans. The mixture of fascination and fear, which often gripped whites when they turned their attention to blacks, had a long historical tradition about which South Side entrepreneurs could scarcely have been ignorant. The Sporting Set was probably beyond this stage of racial perception, but many white tourists in Chicago must have responded to it. The lyrics to Fred Fisher's famous popular song "Chicago That Todd'ling Town" gaily inform listeners that:

> More Colored people up in State Street you can see,
> Than you'll see in Louisiana or Tennessee.[29]

For curious whites, the expressive dimensions of the new northern urban African-American culture could be experienced in the black-and-tan cabarets—for a price.

Inevitably, a new generation of entertainment entrepreneurs arose to satisfy the demand for urban styles of recreation and personal expressiveness. Jazz became an influential force in this development.[30] In the commercialized bright-light districts on the North and West Sides of Chicago, large dance halls were constructed during the nineteen teens and twenties, but in the black ghetto, entertainment focused on more limited capital investments like night clubs and cabarets. As the historian of black Chicago, Allan H. Spear, points out: "Negroes were completely excluded from most commercial amusements—skating rinks, dance halls, [night clubs], and amusement parks." The one amusement park for blacks— Joyland Park at Thirty-third Street and Wabash—never became a major force in South Side musical enterprise.[31] Given the lack of city financed recreational facilities, a larger role was played in young people's leisure time plans and employment opportunities by cabarets, dance halls, movie theaters, and pool halls.

Until the opening of the Savoy Ballroom at the corner of 47th Street and South Parkway on November 23, 1927, African-Americans had no large, commercial dance hall. An effort made around 1913 to create "a model dance hall" on the South Side foundered on the opposition of whites, who feared that it would attract "vicious" elements. White City Ballroom on 63rd and Cottage Grove, outside the ghetto of the twenties, catered to whites, although its owners, the Beifields, regularly employed black orchestras. The Trianon Ballroom at 62nd and Cottage Grove was similarly reserved for whites. Black youth tended to concentrate on what was available to them, vaudeville theaters and cabarets.[32]

Chicago's most widely discussed Jazz Age cabarets were the "black-and-tans," clubs located in the South Side ghetto which presented entertainment by black performers and catered to both blacks and whites. The phrase stemmed from a slang expression used after the Civil War to describe those Republican factions that included both blacks and whites.

In Chicago, "black-and-tan" indicated a night club in which blacks and whites could interact with one another in certain socially stylized ways, talking, flirting, drinking, dancing, and listening to music. Such activities might lead to much more intimate social contacts thereafter, but those more explicitly sexual relations did not take place in the black-and-tan cafes.[33]

The black-and-tans pursued differing policies according to a variety of factors. Some black-owned cabarets, usually not the largest, would admit whites but catered overwhelmingly to black customers, who thought of these clubs as neighborhood institutions. Other clubs, usually larger, were owned by blacks and designed, in part at least, for an interracial clientele. A few of the largest South Side clubs were owned by whites who hoped to attract a primarily white audience with African-American entertainment. Some cabarets—such as the Pekin Inn, as we have seen—changed their racial policies according to economic and political pressures, so that a club known for catering to whites at one moment might at another turn to a more African-American emphasis.

Before the construction of the Grand Terrace Cafe in 1928, Chicago seems to have had no black-and-tan cabaret like Harlem's Cotton Club, where, from the start, only whites were allowed as customers, although the high prices at the South Side's Sunset Cafe effectively excluded most South Siders. Chicago's relatively freer interracial mingling in the clubs was an essential part of its reputation as a more earthy, elemental jazz city than New York, but much depended upon the particular establishment and at what particular point in its history. As Harlem stride pianist "Willie the Lion" Smith put it after playing for several months in 1923 at Chicago's Fiume Cafe, "... there was a lot more mixing of the races in Chicago at that time than there was in New York. I sure found the 'toddlin' town' to be real friendly."[34]

During and just after the war, increasing numbers of both black and white entrepreneurs, responding to black immigration and to white Chicago's growing awareness of what the Chicago *Tribune,* the city's leading newspaper, called "the incoming hordes of Negroes," opened cabarets on the South Side. Several of these newer clubs became important foundations of Chicago jazz: the DeLuxe Cafe, Dreamland Cafe, the Royal Gardens Cafe, the Apex Club, the Plantation Cafe, and the Sunset Cafe.

The most frequently discussed and promoted in the black newspapers was William Bottoms' Dreamland Cafe at 3520 South State Street.

On October 7, 1914, this "new and magnificent hall" had a grand opening to advertise its 18 electric blow fans, 5 exhaust fans, 125 electric lights,
and its 800-person capacity dance floor whose boards were laid in a circle.
Dreamland later reopened under Bottoms' ownership in May 1917; "an
orchestra on an elevated stage or platform discourses fine music throughout the afternoon and evenings." Even the old-fashioned Chicago *Broad
Ax* declared the Dreamland Cafe "one of the most pleasant places of
amusement on the South Side." In June 1917 its "Original Jazz Band"
was one of the very first on the south side to spell "jazz" with two z's. In
September 1918, the Dreamland was reported to have "the best ragtime
band in Chicago, accompanied by Bertha Hall, Alberta Hunter, and Mr.
Tom Mills . . . The New Orleans Jazz Band hits the rail[s] all the time at a
high rate of speed . . . Bottoms really understands how to provide first
class and 'catchy' amusement." In 1919, Bottoms hired Joseph "King"
Oliver and his band away from the DeLuxe Cafe. From that point
forward, Dreamland remained in the forefront of jazz development in
Chicago, particularly in 1925–26, when Bottoms featured Louis Armstrong in Lil Hardin-Armstrong's Dreamland Syncopators.[35]

Bottoms enjoyed the sort of admiration and neighborhood support
which the ghetto community had extended to Robert Motts and Teenan
Jones. Although Bottoms is nowhere mentioned as a leading politician,
he may have used his club the way Motts and Jones used theirs. But
whenever any black entertainment entrepreneur invested his money on
the South Side, hired neighborhood men and women to work in his
establishment, and featured black entertainers on a regular basis, he
championed the racial ambitions of South Side Chicago even when
many of his customers were white. White cabaret entrepreneurs, on the
other hand, could hire black entertainers and service personnel in their
cabarets, but their own racial identity and that of their white clients
placed the African-Americans in a secondary and more subservient position.[36]

Although the South Side supported any clubs and cabarets that employed black musicians, entertainers, and service personnel, whole-
hearted loyalty was reserved for clubs like Dreamland, which the Chicago *Defender* called "a first class resort owned by a member of the Race."
In the black-and-tan business, the race of the owner-proprietor could
affect the taste with which black entertainment was presented to the
general public and the courtesy with which black customers were received. White proprietors and managers were more likely than black

ones to fall into the sort of racial stereotyping which traditionally resulted from the white commercial regulation of black musical entertainment. As the *Defender* put it concerning Dreamland:

> There are a great many of our people who refuse to patronize caba-
> ret and entertaining parlors operated by others than our own
> folks.... The members of the Race who patronize cabarets have al-
> ways taken pride in this particular resort as operated by Billy Bot-
> toms. Residents and business men of the Race throughout the city
> could feel safe in taking their close friends and the members of
> their families there with the knowledge that nothing would be al-
> lowed, by word or act, to cause complaint.37

The *Defender* minced no words in its condemnation of the South Side greed and complicity in white racist elements of the cabaret scene. "Scores of young men," taking their cues from the neighboring Levee vice district, cultivated "a profitable field in acting as escorts to young white men slumming through the south side." Such "fixers," also re-ferred to as "go-betweens" in white urban reform literature, guided groups of white males from cabaret to cabaret and located afterhours "buffet flats," apartments where food, drink, piano music, and prostitutes awaited them. Black waiters, who acted as "fixers" in white-owned caba-rets, embarrassed respectable black customers, who thus relied on "de-cent," "orderly," "race" cabarets. Bottoms encouraged neighborhood pride by making Dreamland the sort of club of which the community could be proud.38

The second leading South Side site for jazz music was the Royal Gardens Cafe at 459 East 31st Street. Sometime before 1918, two police-men had been killed on the premises, and the City Council had decreed that no further cabarets might be opened there. In exchange for a one-third interest in a proposed lease on the premises, Louis B. Anderson, elected Alderman of the Second Ward in 1918, managed to get this restriction removed and joined with Dreamland Cafe's William Bottoms and Virgil Williams as equal co-owner of the Royal Gardens Cafe. When Mrs. Florence Majors obtained title to the building in 1921, Williams, Bottoms, and Anderson sold out, and the Royal Gardens Cafe was re-named the Lincoln Gardens.39

The Royal Gardens Cafe was in some ways a different sort of institu-tion from Bottoms' Dreamland Cafe. It was the largest dance hall on the

South Side, until the Savoy Ballroom was constructed in 1927. Virgil
Williams, a director of the Liberty Life Insurance Company and presi-
dent of the Royal Gardens Moving Picture Company, directed the Royal
Gardens, which many members of the community regarded as a commu-
nity youth institution. Drummer Warren "Baby" Dodds insisted that
Royal Gardens/Lincoln Gardens was not a "black-and-tan," as the New
York *Clipper* had claimed, but rather a neighborhood dance hall. The
Juvenile Protective Association, an influential white urban reform orga-
nization, agreed that the facility was essential to youth culture on the
South Side and kept paid chaperones on the dance floor. The JPA even
went to federal court to defend the dance hall from an official attempt to
close it down.

Although whites were admitted, white Chicago jazzmen Eddie Condon, Bud Freeman, and George Wettling noticed that they were not
given a particularly warm welcome when they went to listen to King
Oliver's Jazz Band at the Lincoln Gardens. Wettling, however, described
the dance hall this way:

> There was a painted canvas sign about two by four feet square
> hanging outside the best-looking building that housed the Lincoln
> Gardens Cafe, a sign that read "KING OLIVER AND HIS CREOLE
> JAZZ BAND."... The thing that hit your eye once you got into the
> hall was a big crystal ball that was made of small pieces of reflect-
> ing glass and hung over the center of the dance floor. A couple of
> spotlights shone on the big ball as it turned and threw reflected
> spots of light all over the room and the dancers. Usually they'd
> dance the Bunny Hug to a slow blues like *London Blues* or some
> other tune in a like slow blues tempo, and how the dancers would
> grind away. The ceiling of the place was made lower than it actu-
> ally was by chicken wire that was stretched out, and over the wire
> were spread great bunches of artificial maple leaves.

Tenor saxophonist Bud Freemen remembered that on one occasion, at
least, there "was a man on the dance floor who would direct the dancing;
if he saw people who were dancing too close to each other or staying in
one spot he would say: 'Get off that dime, man. Let's move it around.'"[40]

While Virgil Williams was its proprietor and James Griffin its manager from 1918 to 1921, the Royal Gardens consistently hired Clarence
Muse, a mainstay of black vaudeville who went on to a long career as a
character actor in Hollywood, to stage what were often elaborate floor

shows. The facility led the way into the jazz age by featuring King Oliver's Band for an exceptionally long run, from 1919 to 1924 (with time out for a trip to the West Coast). It caught fire under suspicious circumstances on Christmas Eve 1924 and reopened only briefly, in 1927, as the Charleston Cafe and then the Cafe de Paris, before reverting to its original name in 1928.[41] Like Dreamland, this cabaret, even under white ownership, served the local community by providing one of two relatively large dance floors in a neighborhood otherwise deprived.

The Sunset Cafe, opened by Edward Fox and Sam Rifas on August 3, 1921, at the corner of 35th and Calumet, led the white-owned black-and-tan cabarets on the South Side. This cabaret was subsequently managed by Joe Glaser, a white, who later became Louis Armstrong's manager. The club hired Percy Venable, one of the two leading black floor show directors, to design and stage its floor shows; and Glaser consistently featured black bands—those of Art Sims, Sammy Stewart, and Carroll Dickerson among them—and helped Louis Armstrong further his career as an entertainer of white audiences.

The Sunset, later the New Grand Terrace, was the most widely known black-and-tan to which crowds of slumming whites ventured when the uptown theaters closed. When it first opened, anonymous leaflets, allegedly written by "The Citizens' Betterment Club," circulated through the South Side claiming that this new white-owned club presented "immoral" entertainment "of the lowest type." Co-owner Edward Fox angrily refuted the charges in the *Defender,* insisting that he ran the Sunset "in a clean, businesslike and legitimate manner," and reminded the readers that the Sunset Cafe employed "upward of fifty people." Black musicians who worked there and white musicians who sat in with them concur that nearly all of the customers were white.[42]

The second leading white-owned black-and-tan cabaret on Chicago's South Side was the Plantation Cafe, located at 338 East 35th Street near Calumet Avenue, diagonally across the street from the Sunset Cafe. The Plantation was initially owned by Edward Fox and Al Turner, and, despite the minstrel show imagery in the club's name, its interior decorations, and some of its entertainment, it presented Joe Oliver's Dixie Syncopators, an important jazz band, from February 1925 until the club was repeatedly bombed in the spring of 1927.

The Dixie Syncopators featured such influential black jazz instrumentalists as reedmen Barney Bigard, Albert Nicholas, Paul "Stump" Evans, and Darnell Howard, trombonist Edward "Kid" Ory, and drummer Paul Barbarin. *Variety* reported that Oliver's band dispensed "real jazz," "loud, wailing, and pulsating," jazz with "no conscience," to a house packed with whites—sophisticated high school youngsters, office clerks, and out-of-town businessmen who to get in paid a dollar on weekends and fifty cents during the week.

The early history of the Plantation Cafe, told here for the first time, reveals important information about the elaborate arrangements which kept gangster-owned jazz speakeasies in business during both national Prohibition and the tenure of Mayor William Dever, staunch defender of prohibition. The thrill of illegal drinking in public places that were likely to be raided by the police added another touch of daring to the cabaret experience without exposing the customers to any serious confrontation with the law. The Plantation Cafe had been Al Tierney's Auto Inn, a particularly notorious, segregated, whites-only cabaret where Dapper Dan McCarthy had shot Steve Kelleher to death. Shortly after Dever's election as mayor in 1923, he closed the Auto Inn, and Tierney moved further south into the all-white Woodlawn rooming house neighborhood and opened the Pershing Palace in the New Pershing Hotel at 61st and Cottage Grove Avenue. He continued to own the building at 35th and Calumet.

The Plantation Cafe went into business in 1924 without a license and quickly established itself as a wide open, all night black-and-tan, allegedly controlled by the Capone syndicate. *Variety,* reflecting continued reform pressures, emphasized the prevalence of "undesirable characters" at the Plantation Cafe, inside and out; it said that "white women" were not safe in the club, due presumably to gangsters, since most of the blacks were out in the street selling pints of gin (asking price $3) and "bonded" bourbon (asking $8). When federal prohibition officers seized a pint of whiskey which had allegedly been passed from a black to a white man in the Plantation Cafe on the night of November 9, 1924, the agents took it back to their laboratories to verify that the liquid was (indeed) alcohol. Twelve days later, their suspicions scientifically confirmed through laboratory analysis, the officers raided the Plantation, searched unsuccessfully for more alcohol, and arrested manager John Paley, who was subsequently released, since he had not witnessed the transaction in question, the waiter involved had disappeared, and no restaurant bill existed to document the sale of any alcohol.[43]

Despite obvious illegal activities and much discussed underworld connections, downtown newspapers accepted publicity from clubs like the Plantation Cafe, and uptown Loop theater programs carried ads for the Apex Club, an elegant after-hours spot on the second floor just across Prairie Avenue from the Chicago *Defender*. The Apex, formerly called Club Alvadere and The Nest, was owned by boxer Joe Louis' manager and backer Julian Black. It presented stylistically innovative jazz as early as 1926 by Jimmie Noone's orchestra, featuring pianist Earl Hines. The club catered to a wealthy white clientele, and many of the white jazzmen who called themselves "the Chicagoans" frequented the Apex Club in order to listen carefully to Noone's band.[44]

The South Side's famous cabarets, influential channels for interracial contacts in a segregated city, served to shape patterns of commercialized entertainment and personal expressiveness into emotionally powerful interracial rituals. Plying their trade on the northern edges of the black ghetto, an area associated with, if not in, the old Levee vice district where alcohol, drugs, and prostitution had been available for many years, cabarets saw their share of what were then called "vicious" activities like prostitution. What one black newspaper called "the white plague"—white men in search of black women and white female prostitutes looking for business among black men—gravitated to the Stroll. In 1916, black citizens had rallied against an attempt by a white North Side sporting set entrepreneur to turn the area around 31st Street into a red light district.[45] It seems safe to assume that interracial prostitution flourished around the entertainment district of the South Side.

But, with the exception of the spring of 1923, when the police closed the bordellos in the old vice district and forced prostitutes to solicit openly in the cabarets, the most assiduous of the cabaret investigators—Jessie Binford, Paul M. Kinzie, and Nels Anderson of the Juvenile Protective Association—found some prostitution, but less than they had anticipated. According to their reports on the Dreamland Cafe, female prostitutes of both races, seated at tables, solicited by winking at men who passed by. Prostitutes hung around the door to the Royal Gardens Cafe. (One told a JPA investigator to "go-to-hell," when he plied her with more questions than drinks.) Prostitutes or their "fixers" worked the sidewalks outside the clubs.

But investigators working for the JPA and the associated American Social Hygiene Association, who entered cabarets during the early morning hours and remained only long enough to take notes, reported that such prostitution, while closely related to the black-and-tan sensibilities, was not the main business of the cabarets known for their jazz music.

While prostitution seemed to them to be the integral ingredient at the Entertainer's Cafe at 209 East 35th Street, the Pekin Inn during its declining years of short-lived white ownerships, the Edelweiss Gardens, the Pioneer Cabaret, and the Schiller Cafe, to their surprise, it did not provide the focus of cabaret activities. Prostitutes were observed to work the bars in the cabarets known for their jazz orchestras, but the clubs provided them with no special facilities or consideration. White slummers did not go into the leading jazz-oriented cabarets to sleep with prostitutes.[46]

To the most morally exacting observers, Chicago's jazz clubs of the twenties seemed to be a prominent feature of "vicious" neighborhoods but one that created a separate, specialized world of musical entertainment. Instead of "vice" itself, the Dreamland, Sunset, and Royal Gardens specialized in presenting unprecedented spectacles of interracial contacts in social dancing. A major attraction for many white customers was the experience of Anglo-Americans and African-Americans dancing together to jazz music. Both paid entertainers and paying customers joined in this social dance chiaroscuro. Following the instructions of a master of ceremonies, customers would seat themselves at small tables and watch a floor show revue in which African-American vocalists and dancers performed what reform investigators considered to be "suggestive songs" and "improper," "indecent," "contortions and jazz dances."

Jazz musicians provided the music for these shows and, once the customers "were permitted to dance," played dance music for them as well. Out on the crowded floor, hard by the slightly raised "stage," the customers, often mixing with prostitutes, imitated the steps, gestures, and movements which had been performed for them. Many customers, particularly those who appeared to be wealthy Gold Coast parties "slumming" in the ghetto, took a keen interest in this inter-racial dancing. In the clubs which preserved racially segregated seating arrangements, this inter-racial dancing must have generated a voyeuristic appeal to those who remained at their tables. Paul M. Kinzie, on the trail of illicit vice on the evening of December 10, 1923, and generalizing about his observations, felt that "slumming parties ... are apparently pleased with the *atmosphere of sensuality* and *find delight in seeing the intermingling of the races.*"[47] Black-and-tan cabarets sold not vice but suggestive African-American musical entertainment which helped customers create an atmosphere of inter-racial "sensuality."

In part, the market for this entertainment ritual grew from the perception among late nineteenth- and early twentieth-century middle- and upper middle-class whites that the nervous pressures of modern urban life required release and relaxation from moral and intellectual strenuosity. White "antimodernists," as described by T. J. Jackson Lears, hoped to learn from "'Oriental people, the inhabitants of the tropics, and the colored peoples generally.'"[48]

But this stylized "black-and-tan" process took on greater ritualistic and symbolic power in the atmosphere of inter-racial fear and hatred expressed in Chicago's murderous race riot of 1919. Twenties jazz and the black-and-tans involved what John Szwed has called "the embodiment of culture," in which Caucasians, depending upon their interests, temporarily adopted the dance and/or the musical mannerisms of African-Americans, taking on, in their own minds, the characteristics of South Side Chicagoans in ritualized, often racially stereotyped, song and dance. The musical rituals of the black-and-tan cabarets were *inclusory* and served to explain and interpret in an essentially reassuring manner the presence of other, exotic, and perhaps dangerous peoples in their midst. Polyphonic jazz in the black-and-tan cabarets of twenties Chicago suggested some of the disorder and moral confusion of contemporary city life, but it also provided reassuring harmonic and rhythmic structures that helped customers to allay their fears of modern life and interracial violence. Musicians and musical entrepreneurs quickly learned how to earn money by staging elements of the popular night-life fantasies cherished by white customers.[49]

At least some of the jazz cabarets turned away from the black-and-tan business toward their other, more immediate market, becoming, at certain times, black neighborhood institutions. Theoretically, of course, white customers might walk in at any time; but, in practice, they tended to come late at night, the traditional association of night-time with "darktown" holding strong in the white sporting set. Thus, South Side jazz cabarets acted as black-and-tans from the late evening through the early morning hours.

Under mandatory police closing orders brought on by World War I but abandoned thereafter, customers had to leave by 1:00 a.m.[50] and tended to seek after-hours clubs like the Book Store and the Apex Club in order to extend the evening. But the Royal Gardens and Dreamland Cafe advertised a variety of alternate schedules and special events for the local community: matinees from 3:00 to 7:00 p.m. on selected weekdays caught members of the neighborhood coming home from work; "Blue

Monday" specials lowered prices; breakfast dances from 4:00 a.m. to
midmorning on Sundays became a standard feature of the jazz age;
special gatherings for the musicians' union (Local 208) and the waiters'
union encouraged ghetto institutions; benefits, like that staged by the
Plantation Cafe for the dependent relatives of soldiers killed at Camp
Grant during the encampment of the all-black Eighth Regiment in 1925,
further tied the black-and-tans to the local community. The Chicago
chapter of the National Association for the Advancement of Colored
People met at the Dreamland Cafe in June 1926.[51]

But the jazz cabarets were highly controversial within the black as
well as the white community. Black religious leaders and old time South
Siders with the sort of Victorian attitudes which appeared in the *Broad
Ax,* disapproved of dance halls, cabarets, ragtime, jazz, leisure time license, and, in fact, the whole "bright-light" transformation of their
neighborhood. Dr. M. A. Majors, editorial writer for the *Broad Ax,*
roundly condemned all of the latest entertainment trends, especially
"ugly, low, nasty dances" and the gangs of young black toughs who hung
around the pool halls "armed to the teeth."[52] Despite such opposition, the
Broad Ax did support and promote the black owners of entertainment
enterprises.

On the other hand, the Chicago *Defender,* which had played such an
important role in stimulating the Great Migration, easily adjusted to the
new social patterns of life on the South Side, adopting the new journal-
ism, and encouraging black-owned cafes and dance halls between 26th
and 39th streets, just as it promoted black-owned grocery stores and fish
markets. The *Defender* considered Motts' Pekin Inn, the two Elites, the
DeLuxe Cafe, Dreamland, and the Grand Theater to be "up-to-date,"
"Afro-American" establishments that enjoyed "phenomenal patronage
by our people"; where every employee was "a member of our race," and
where "customers would always receive respectful attention."[53]

While not, perhaps, an unmitigated moral blessing to the community,
night-club investments and the music they presented expressed the kind
of optimistic entrepreneurial spirit characteristic of black Chicago in the
1920s. The *Defender* celebrated the profits to be made in "supplying the
needs of a cosmopolitan population"; banking, insurance and real estate
all found "pre-eminent opportunity and success in Chicago."

... Chicago is the embodiment of the dream of Booker T. Washington.

Chicago takes pride in its artistic temperament ... Along the lines of music there is a generous and growing development that have [*sic*] given the community a fineness of taste.54

Jazz and business did not seem to be contradictory activities, but rather were integrally related to one another and to the ideal of success. As *Defender* columnist Dave Peyton put it:

> Chicago musicians are away [*sic*] ahead of musicians of our group in other cities of the country. Their achievements have been wonderful.... Let us make the world respect us. Ours is an art ... Work together, acquire real estate, and then you will be independent.55

Musical enterprise was just one important element in a broader and more varied effort to find economic opportunity in commercializing various forms of popular amusement. From the earliest days, for example, Robert Motts and Teenan Jones combined their musical entertainment with vaudeville, legitimate theater, gambling, and such fast growing sports as boxing. Teenan Jones headed the reception committee for boxing champion Jack Johnson when he returned to Chicago from his celebrated victory over Jim Jeffries in Reno, Nevada, in 1910. Dreamland Cafe owner Bill Bottoms later promoted Johnson in a series of exhibition bouts in the Dreamland basement when the ill-fated black champion again returned to Chicago after his release from Leavenworth prison.56

Jones and Bottoms involved themselves in efforts to turn the all-black American Giants baseball team into a commercial proposition. Both cabaret owners contributed their facilities to the team for banquets at the start of the season. Bottoms and William S. Abbott jointly promoted at least one football game in 1922 between an all-white and an all-black team, a highly publicized match won 7–0 by the blacks. Jazz was born and matured in the years when official "sporting life" evolved into commercial popular culture on the South Side.57

Most revealing in this connection were the efforts of owner and manager Virgil Williams of the Royal Gardens to organize an all-black "Royal Gardens Motion Picture Company" to train blacks in movie making and photoplay production. By 1920, Williams had produced three films, the third, a comedy, earning $10,000 in bookings. This company, like so many other black entertainment initiatives, did not survive.58

The troubled history of Chicago's Joyland Amusement Park, an

African-American enterprise at 33rd and Wabash reflected the ambitions which nearly all South Side entrepreneurs shared for an expanding commercialization of entertainment. In 1923, South Side Attorney Augustus L. Williams spearheaded an attempt by "many of the best colored men and women in this city," including Virgil Williams, editor Robert S. Abbott, and *Broad Ax* editor Julius F. Taylor, to open an amusement park for blacks, a pressing need since all of the large Chicago amusement parks were racially segregated. Moreover, many of Joyland Park's backers saw the amusement park's merry-go-round, ferris wheel, whip, and Venetian swing as healthy alternatives to cabarets:

> ... thoughtful citizens ... believe that the People should have outdoor amusement, rather than to live and pass their time in Cabarets, stuffy indecent picture and vaudeville houses which tends [sic] to create criminals.

Yet when Joyland's backers applied to the City Council for a license to operate an amusement park, Second Ward Alderman Louis B. Anderson, with support from Republican Ward Committeeman Edward H. Wright, introduced a Council order to close the facility. Mayor Dever vetoed the order and Joyland remained open, but it continued to be plagued by opposition from the cabaret owners and their political representatives in City Hall.[59]

The influence of South Side politicians at City Hall directly connected African-American musical enterprise to white Chicago, creating political protection for Chicago's black-and-tan speakeasies, and encouraging an inter-racial audience for jazz. William Hale Thompson, the Republican candidate for mayor in 1915, and one of the most flamboyant of America's urban machine politicians, appealed, both publicly and privately, for the black vote. He worked through white South Side lieutenants like George F. Harding, the millionaire alderman of the Second Ward, for support of the black entrepreneur/politicians. Harding, a crucial intermediary between City Hall and the South Side's cabaret politicians, was a real estate magnate with offices at 31st and Cottage Grove Avenue; he reportedly owned hundreds of buildings on the South Side and charged his black tenants reasonable rents. In return for the overwhelming support of black voters in his successful campaigns of 1915, 1919, and 1927, Thompson appointed many blacks to city hall positions.

Thompson, a Nebraska cowboy, dedicated sportsman, and two-fisted bourbon drinker, personified Chicago's new, open affair with the entrepreneurs of urban show business. His political coalition of Irish, German, Italian, and African-American minorities included most of Chicago's important white and black night-club entrepreneurs. Big Bill's campaigns were part tent show, part circus, and all show business, as he stumped the city with vaudeville acts, presenting vocalists like Sophie Tucker, and black jazzbands. He was known to shout "Get a horn and blow loud for Chicago ... Put on a big party! Let the jazz band play! Let's show 'em we're all live ones!'"[60]

In 1919, the second of the three mayoral election years that witnessed Thompson's triumph in Chicago, Virgil Williams donated his dance hall for a "Big Second Ward Republican Harmony Dinner" on December 4th. With a $3000 check contributed by George Harding, Williams organized an elaborate banquet for leading white and black politicians. Mayor Thompson spoke, "pressed the flesh," and listened to "the Royal Gardens Orchestra and its entertainers." When Big Bill Thompson landed the Republican National Convention in Chicago in 1919, Virgil Williams turned the Royal Gardens into an entertainment center for convention delegates.[61]

In return for the organized political support constructed through South Side entertainment establishments, Big Bill Thompson's political machine protected speakeasies, with which jazz, of course, was closely associated. Thompson was caught, on the one hand, between the demands of socially influential urban reformers and temperance advocates for total repression of cabarets and, on the other, the more directly persuasive political support of night-club entrepreneurs and bootlegging gangsters. His solution appears to have involved police surveillance and periodic raids of establishments that flagrantly violated any of the city ordinances governing their operation, while leaving the majority of the clubs alone. Throughout his three terms in office, Thompson retained a special mayoral prerogative to issue licenses to any club he wished, regardless of contrary political pressures in City Council. Interference with the jazz cabarets was left to federal agents, who were seriously overburdened. The 134 federal Prohibition agents working out of Chicago had to cover all of Illinois, Iowa, and eastern Wisconsin. Throughout the 1920s, selected South Side clubs were periodically raided and closed, but they usually reopened. In 1922, for example, Izzy Shorr's Entertainers' Cafe was closed for one year for violations of the Volstead Act. Nothing came of a simultaneous attempt by the prosecution to have the club's jazz music, itself, declared morally dangerous. Even Dreamland Cafe was closed for a time late in 1923 and early in 1924.[62]

Musicians and entertainers who made their living in cabarets had become inured to this game. In 1917, vaudevillian Billy King and his company had staged a "screaming farce" at the Grand Theater about a cabaret where the show proceeded merrily "until a copper showed up at the back door ...":

> Table cloths were then raised and hanging from the edge of the table was a Sunday school motto, and Billy King, who was enacting the part of the waiter, transformed himself into a minister and continued exhortations.

Ten years later pianist Earl Hines recalled that he and the other entertainers at the Sunset Cafe would have to crowd into a patrol wagon and ride to the police station nightly. "I stood up in the wagon so often on those trips, I finally decided to *run* and get a seat when the police came."[63]

The continuing threat of police raids, far from dampening public enthusiasm for South Side night life, seemed to contribute just the right note of excitement to Chicago's jazz scene, mixing with new styles of personal liberation—clothes, insiders' slang, cigarettes, bootleg gin, marijuana (called "gage"), sexual expressiveness, and interracial mingling—to add drama to the new music. In fact, the notoriety gained in defying the reformers and fending off court injunctions merely served to make cabaret entrepreneurs like Teenan Jones, Bill Bottoms, Ike Bloom, and Bert Kelly into "personalities" customers flocked to see at close range.[64]

In 1923, prohibitionist Democrat William Dever won the mayoral election, but Thompson would win back City Hall in 1927. In the interim, cabarets like the Sunset Cafe and Dreamland, while having to adjust, ultimately continued much as usual. To square accounts with those in the Republican party who had dropped him from the ticket, Thompson delivered his usual blocs of votes to Dever. Once in power, Dever closed a number of clubs, including Tierney's Auto Inn, located at the center of the most respectable black neighborhood and catering to whites only. He also closed Dreamland, but this unleashed such an outcry among Bottoms' many South Side friends that it soon reopened. Of the South Side cabarets, the Entertainers Cafe was the biggest loser to Mayor

Dever's reform activities; it was closed in 1924 for a year and a day and reopened for only a brief time thereafter.[65]

Dever became famous for his "beer war" against the "soft drink palaces" that secretly sold beer. Up until 1923, an unspoken gentlemen's agreement had held that prohibition applied only to hard liquors. By November of that year, the Democratic mayor had revoked the licenses of 1600 businesses and 4,031 saloons. Most of them reopened, due to the many legal ambiguities in the enforcement of the Volstead Act; the courts issued restraining orders prohibiting further interference with such businesses. During the Dever regime, moreover, 182 cabarets were licensed.[66]

When Big Bill swept back into office in 1927, the musical entertainers turned out in force to honor the mayor and South Side Aldermen Louis B. Anderson and Daniel Jackson at a Victory Ball held on June 9th at the Eighth Regimental Armory. The entire troupes from the floor shows of the Cafe de Paris (the old Lincoln Gardens) and the Sunset Cafe appeared, the latter putting on "a thunderous show, speed and plenty of it. They danced, they black bottomed and did everything else to make merry...." Louis Armstrong and his orchestra, representing the Sunset Cafe, thrilled the crowd.[67]

Mayor Thompson's encouragement of bright-light enterprise facilitated the determined efforts of black musicians to break through racial barriers around the South Side and secure regular, contracted playing jobs in white Chicago. This was never easy, due to the determined opposition of the all-white Local 10 of the American Federation of Musicians, but black band leaders did successfully penetrate white Chicago, and even secured several long-run engagements at some prominent night-life institutions. Sometime between 1915 and 1916, George Filhe led a group that included Charles Elgar at the Fountain Inn at 63rd and Halsted, while another instrumental ragtime group associated with Elgar, one which included trumpeter Manuel Perez, clarinetist Lorenzo Tio, Jr., trombonist Ed Atkins, and drummer Louis Cottrell, Sr., had played sometime soon thereafter at Mike Fritzel's Arsonia Cafe at 1654 Madison Street.[68]

In 1916, an Elgar group had been the first black orchestra to play a long-term engagement in a white dance hall on the West Side—Paddy Harmon's Dreamland Ballroom—performing both "rough dance music for mostly Jewish people and Italians" and "sweet waltzes" whenever urban reformer Jessie Binford dropped in to take the moral temperature.

This cavernous old one-story, barn-like building under the elevated tracks at Paulina and Van Buren streets also featured Charles "Doc" Cook's Orchestra, another top black dance band, for a long run from 1922 to 1927, and Joe Oliver's band from time to time. Cook's large orchestra also played for several seasons at the city's Municipal Pier and at white South Side's appropriately named White City Ballroom.[69]

Before the Great Migration of Deep South blacks to Chicago it was not unusual to find African-American bands playing in all-white clubs. Among other things, race orchestras seemed to offer greater expressiveness. In February 1916, for example, a reporter from the Chicago *Herald* visited the Cafe de l'Abbe in the Hotel Normandy at North Clark and West Randolph streets and found a black instrumental quartet led by an unnamed violinist [Elgar?] playing upstairs. Using what were to become metaphorical jazz clichés, he tried to describe the music's impact on a young, innocent girl, accompanied by a much older, more sophisticated man:

> Upstairs was hidden away a quartet of musicians. They were black,
> but were far from being in mourning. The violinist had taken his
> instrument, the most wonderful interpretive instrument in the
> world, and he was making it talk. Here's what that violin said [to
> "Mary," the simple innocent girl] and the cocktail added its aban-
> doned echo... "Ah, Mary, isn't it wonderful to be loved..."

The music jumped from dreamy romance to "Ragtime Pipers of Pan," as the spotlight suddenly illuminated two young female entertainers.[70]

Throughout the 1920s, black band leaders who could supply refined and regimented music were those who most often were able to secure longer running, contracted jobs in all-white clubs and dance halls. Gold Coasters, for example, provided a lively market for society dance bands such as that of William Samuels. Black musicians often worked for private parties, particularly on New Year's Eve, but there were one-night stands, not the sort of regular work on which to build a career.

With the help of cabaret entrepreneur and loyal Thompsonite John M. Kantor, Eddie South, "The Dark Angel of the Violin," and black bandleaders Albert Wynn and Jimmy Wade led orchestras at Kantor's elegant Moulin Rouge Cafe at 416 South Wabash Avenue in the Loop. South, a classically trained violinist, played a soft, refined style of jazz indicative of the more "legitimate," familiar sounds which white Chicago expected of black orchestras outside of the black-and-tans.[71]

Even black jazz bands which presented a studied New Orleans polyphonic style worked for many years at Bert Kelly's Stables at 431 N. Rush Street in the Towertown bohemian section of town. This cabaret advertised itself as having brought jazz to Chicago in the first place. Kelly presented bands led by clarinetist Yellow Nunez, King Oliver (for a three-month engagement starting in November 1924) and Freddie Keppard, "the World's Greatest Colored Jazz Cornetist, formerly at Purcell's on San Francisco's Barbary Coast." Keppard's band played a long run at Kelly's Stables before the cornetist's fabled drinking habit forced his clarinet player, Johnny Dodds, to take over the group in 1926. Dodds remained in charge of the band, which included his brother Warren, Honore Dutrey, and Charlie Alexander, until 1932.[72]

Chicago had its North and West Side clubs and dance halls where black orchestras entertained white audiences, but racial integration stopped there. Some influential clubs never hired black musicians, and, of course, blacks were not allowed into most of these clubs as anything other than employees. Whites could venture into black cabarets, but blacks could not enter most night clubs or dance halls in white Chicago, not even those white establishments where black orchestras provided the music. As a result, many, perhaps the majority, of white Chicagoans, both musicians and customers, experienced the jazz age in all-white environments; only an inner circle of white musicians realized that the latest innovations in jazz performance were to be found on the South Side. As white cornetist Muggsy Spanier put it: "White and Negro musicians kept to themselves pretty much; only rarely did they sit in with each other; there were no mixed bands."

During the 1920s, Mike Fritzel's Friars' Inn ("Land of Bohemia Where Good Fellows Get Together"), at 343 S. Wabash in the Loop epitomized the limits of racial integration in Chicago jazz.[73] This club featured the white New Orleans Rhythm Kings for a lengthy run early in the decade, a group that exerted a major musical influence on other budding white Chicago jazzmen. Throughout most of the era, Friars' Inn, which was, of course, off limits to black customers, refused to hire black musicians. When, in September 1926, a black band finally maneuvered into the club, Local No. 10 president James Petrillo tried to force them out by claiming that they, like all black bands, were playing for less than union scale. But President Verona Biggs of the black local No. 208 investigated and discovered that the New Orleans Rhythm Kings had been playing under scale, while the black musicians were receiving scale.[74]

Beginning in 1926, an increasing number of black jazz performers broke through the resistance of the white musicians' union and secured contracted jobs in Loop clubs and hotels. In the early part of that year, two black bands had just moved into the Loop's Valentino Inn when the federal government padlocked it for liquor violations. In 1927, Louis Armstrong temporarily broke racial barriers by leading a band in the Loop's Blackhawk Restaurant, one block from the Chicago Public Library. Until 1928, when white vocalist and shimmie dancer Bee Palmer hired violinist Eddie South to lead her backup band, the College Inn featured the dance music of Isham Jones' all-white orchestra. The hotel also presented hot dance music in its exclusive Bal Tabarin after-hours club.[75]

Although the example of black musicians playing in clubs open to white customers did expand the influence of South Side jazz by providing a discrete reminder of South Side Chicago nightlife, continuing racial barriers—segregated locals of the Chicago musicians' unions and segregated night-life institutions—also fostered the growth of a separate African-American musical movement on the South Side, one with remarkably distinctive musical ingredients, a movement that played a far more prominent role within that neighborhood's cultural traditions than did the white jazz evolving elsewhere. The interplay of African-American musical traditions with Chicago's urban institutions influenced the most creative jazz in the city during the twenties. It is time to turn from South Side jazz institutions to the evolution of South Side Chicago jazz within them.

2

The Evolution of South Side Chicago Jazz

THE BLACK MIGRANT musicians who gathered on Chicago's South Side Stroll after World War I swiftly cultivated their distinctive instrumental music, replanting its New Orleans roots in Chicago's highly competitive world of cabaret show business. Here the African-American musical tradition merged with elements of popular and European musical cultures in Chicago to produce jazz styles that responded to their new time and place. A small number of creative musicians swiftly developed and adapted their skills and musical devices so as to enrich and in some cases transform the New Orleans jazz which had migrated north. As there are no recordings of the jazz played in New Orleans before the northern migration, that style cannot be fully compared with the jazz records made in Chicago. And, as the idea of emigration from the South may have appealed primarily to the more ambitious and innovative musicians, at least some jazz in Chicago represented the creatively ambitious side of New Orleans music. But the northern city provided a new cultural context for the immigrants, radiating a powerful cultural symbolism for southern African-Americans in general and for New Orleans musicians in particular. The emigration from New Orleans had taken on something of the spirit of an exodus; Chicago, at least in the beginning, seemed to offer a new start and greater freedom.

Chicago also offered an unprecedented number of jobs for jazzmen and women, work situations that accelerated ongoing patterns of musical evolution; from 1919 to 1929, Chicago jazz became a more varied synthesis of folk music, "lowbrow" minstrel and medicine show entertainment, and "highbrow" musical techniques. Musicologist Gunther Schuller argues that "... the New Orleans style in its pure early form did not survive the 1920s. Even King Oliver and Jelly Roll Morton succumbed to the pressures of changing styles, and their great recordings of that decade represent both the end of an era and the beginnings of a new one." Whether or not such a thing as a "pure" form can be proven to have existed, few would deny that jazz evolved rapidly in northern cities between 1915 and 1930.[1]

Historian Kathy Ogren has shown that professional entertainment networks for black musicians and entertainers in minstrelsy, circuses, tent shows, and medicine shows had begun to take shape on both the national and regional levels before World War I; it would be a mistake, she argues, to interpret black musical entertainment around the turn of the century as "noncommercial folk" music. Much of the evolution in jazz that took place in Chicago during the late teens and twenties, therefore, represented an ongoing cultural process whose fruits leapt to public awareness through Chicago's cabarets, dance halls, and recording studios. As Lawrence W. Levine has put it, jazz, like all black secular music, "manifested the simultaneous acculturation to the outside society and inward-looking, group orientation that was so characteristic of black culture in the twentieth century." African-American musical sensibilities both adapted to and molded changing institutional environments. In moving from New Orleans, for example, some of the social institutions through which jazz had found expression—musical funerals, neighborhood street parades, hayrides, and picnics—were largely left behind. South Side Chicago musical enterprise worked more at adapting transplanted New Orleans jazz to black, black-and-tan, and white cabarets, where "jazz" became an ever more self-conscious act in the structured world of night-club entertainment.[2]

According to banjoist Danny Barker, at least some of the musicians who did not want to take up careers as traveling professional entertainers and therefore remained in New Orleans retained ingredients of pre-professional, communal attitudes toward making music: they played out-of-tune, preferred easy material, and didn't care to improve their instrumental control. According to Edmond Souchon, a lifelong supporter of New Orleans jazz, their "hard-hitting, rough and ready" music was made "without plan from the leaders or sidemen" who sometimes blew "a few bad ones" and dropped out to rest. New Orleans night clubs and dance halls were small, wooden structures where the music was played by a constantly changing group of what Barker called "ham-fat" folk musicians.[3]

Chicago's cabaret floor shows, a central and influential element in South Side clubs, put musicians on display and focused increased attention on visual dimensions of musical performance. Even more than most immigrants to Chicago, musicians learned the new urban standards of manners and personal hygiene advocated by the Chicago *Broad Ax* for all rural immigrants. Jazz musicians, vocalists, and dancers appeared before the black public as models of urban sophistication. South Side musicians unanimously affirmed that the black public held professional cabaret musicians in high regard, particularly as there were relatively few black lawyers and doctors to compete with them for status. The latest in elegant, urbane clothing styles took on increased importance among Chicago's musical entertainers. The stylish black man "was as much a stereotype as the plantation darky," one found both threatening and curiously reassuring by many whites, and also admired by many blacks fresh from the country. A few immigrant jazzmen arrived in Chicago wearing outdated late nineteenth-century clothing styles: tall Stetson hats, tight pants, boxback coats, and high-button shoes. Cornetist Chris Kelly had been known to combine a tuxedo with a blue work shirt and tan shoes when playing in New Orleans. Many of the New Orleans musicians felt lost and lonely in the cold, bleak, and uninviting northern cities.[4]

The more sophisticated style of the 1920s took time to evolve: a few of the pioneer Chicago jazz musicians of the second decade of the century sometimes cultivated styles which appeared gaudy to later arrivals. Jelly Roll Morton had imbedded what he alleged to be a diamond in his front tooth in order to radiate wealth. Some of the so-called Classic Blues singers, who took the entertainment world by storm in 1920, also imbedded diamonds in their front teeth to ensure a glinting smile. Clarinetist George Baquet of the Original Creole Orchestra adopted a theatrical but less flashy diamond horseshoe stick pin in a gaily colored silk cravat.[5]

But Chicago in the twenties was the era of the elegantly tuxedoed jazz musician. Down south, even the parade bands took a casual ap-

proach to uniforms; but, up north, cabarets liked to improve their controversial public image by dressing musicians in tuxedos and encouraging
customers to wear them as well. For their publicity photos, musicians
adopted wing-collared shirts, butterfly bow ties, sharply tailored tuxedos,
gilets, and patent leather shoes. Joe Oliver traded his New Orleans
costume—open-neck white shirt, red undershirt, and suspenders—for a
tux. Louis Armstrong adapted more slowly to new show business demands; his southern habits died hard. In 1922, Armstrong arrived in
Chicago a "green-looking country boy" and two years later still looked
substantially overweight when appearing in New York in "... high top
shoes with hooks in them, and long underwear down to his socks."
Northerners, both black and white, demanded "Up-to-Date," "High
Class Entertainment." As one editorial in the *Defender* put it:

> The World asks, "What can you do?"... we must awaken to the
> fact that we are living in a "show me" age ... "Make Good" is the
> password that opens the door of Success.

Among South Side jazzmen in the 1920s, Earl Hines took the honors for
sophistication when stepping out in his tux, Chesterfield overcoat, and
bowler hat, swinging a walking stick. Hines had been raised in Pittsburgh, where he had learned northern ways. Lacking a union, black
jazzmen in Pittsburgh had dressed up like dandies and hung out on
Wylie Avenue, where bookers regularly hired musicians off the sidewalks. Already, jazz musicians understood clothing as a form of show
business advertising.[6]

   In addition, there were other cabaret bandleaders, both early and late
in Chicago's jazz age, who had lived either in the North or in a border
city before moving to Chicago. Al Wynn, Art Sims, Jimmy Wade, Eddie
South, and Ollie Powers cultivated the sort of personal carriage, social
polish, and verbal skills that prepared them to represent the poorer,
recently arrived southern musicians to band bookers, club managers, and
the public. Wynn and South took bands into white clubs like the Moulin
Rouge up in the Loop. Arthur Sims, whose father Adolph was well
placed in Mayor Thompson's City Hall, possessed what the *Defender*
called "aristocratic" manners and was one of the first jazz band contractors at the Sunset Cafe and Midway Gardens. Fess Williams, who became
Chicago's first black stage band leader, had been educated at Tuskegee
Institute.[7]

Urbane clothes also could signify a sense of personal, economic, and racial pride, particularly since vaudeville had so often dressed the pioneering Creole Orchestra in clichéd theatrical farm clothes. The tuxedo and well-tailored clothes represented a strike against certain kinds of racist stereotypes as well as a personal statement. The *Defender,* for example, printed the story of a physical beating of the all-black Howard's Whispering Orchestra of Gold that was administered by some "crackers" outside of an all-white Miami, Florida, hotel. The musicians "realizing that neatness was an asset ... were appropriately dressed upon every occasion ... This sartorial grandeur did not meet with the approval of the 'crackers.'" Apparently upset by the musicians' neatness, self-assertion, and aloofness, the southern whites allegedly snarled:

> We'll teach you niggers to come here dressed in your white flannels
> and your tweed coats, playing for our dances and looking at our
> pretty white women. Now, go back up North and tell all your nig-
> ger friends.

As musicians appeared before the public as entertainers, they became role models for many younger musicians who either attended their performances or looked in through windows or doors from the streets or alleys. Chicago jazzmen symbolized a new, urban style in racial pride, just as cabarets, cafes, and buffet restaurants signaled that their customers were no longer shabby "Can Toters," trudging home from work with a bucket of beer.[8]

Public cabaret performance in Chicago often demanded a cosmopolitan demeanor. The ambitious movie theater orchestra leader and columnist Dave Peyton never tired of enumerating bad habits which musicians ought to avoid: arriving late, entering the club in a loud, indiscreet manner, smoking and drinking on the bandstand, criticizing the leader, talking unnecessarily and frivolously to one another and to patrons, flirting with customers, playing in a slouched position with legs crossed, and beating time with feet. Peyton often exaggerated, but clarinetist George Baquet similarly recalled that the six members of Buddy Bolden's Band were dozing on the bandstand of the New Orleans Odd Fellows Hall when he first arrived in 1905. In Chicago, Joe Oliver had to discipline trombonist Roy Palmer for falling asleep on the stand. In order to teach his musicians that "the contract depended on being on time," bandleader Earl Hines fined his men $5 per minute late. Some of the

most successful jazz musicians learned important lessons before settling
in Chicago: the Dodds brothers, Pops Foster, banjoist Johnny St. Cyr,
and Louis Armstrong, who had worked on the Streckfus Line's Mississippi River steamboats, brought some invaluable lessons in punctuality,
regimentation, comportment, and memorization with them to Chicago,
benefiting from their prior experience in musical show business.[9]

New patterns of dress and demeanor were just two elements of an
ongoing development of show business calculation and strategies. The
northern nightclub's combination of confined spaces and interracial audiences forced a new attention to show business upon cabaret musicians.
Clarinetist and bandleader Jimmy Noone, a star attraction in the second
floor apartment at 35th and Calumet first called Club Avadere, subsequently The Nest, and finally the Apex Club, was widely known for his
smiling, genial personality. Yet those who knew him well noticed his
anxious attention to the minute details of showmanship, fussing with
signs, lighting, table arrangements, and the like. Noone habitually played
with one eye on the door and sequed into the favorite songs of arriving
fans as they walked into the club.

In the black-and-tans, the mixture of interracial audiences and alcohol also required a wary vigilance. White pianist Tut Soper remembered
going regularly to the Apex Club to hear Noone and his idol, pianist Earl
Hines: "... there'd be drunks coming around and I'd say, 'why don't you
stop bothering the piano player, listen to what he's doing,' and Hines
liked that...."[10]

In step with vaudeville, jazz developed a star system to replace the
folk anonymity of southern musicians. Trumpeter Manual Perez was the
first New Orleans musician to receive star billing in the *Defender;*
Tommy Ladnier, Joe Oliver, Freddie Keppard, Louis Armstrong, and
Reuben Reeves soon followed as cornet and trumpet "Kings." While at
the Royal Gardens Cafe, Oliver saw to it that none of his sidemen
outshone him on the bandstand. Their solos were fewer and shorter than
his own. Jazz commentator Edmond Souchon remembered that, compared with his New Orleans days, the Chicago Joe Oliver was a "star," "a
much more impressive figure now ... 'King,' the most important per-
sonage in the jazz world, surrounded by his own hand-picked galaxy of
sidemen."[11]

But being a "star" also meant being a jazz professional who possessed
endurance, instrumental technique, a distinctive instrumental sound, and
various elements of musical theory. Some New Orleans musicians had

rarely thought of themselves as professionals, living as manual laborers who also played weekend gigs. But in keeping with the increasing specialization of northern industrial society, Chicago jazz, for some musicians at least, became a demanding business which required, among other things, exceptional physical endurance. Joe Oliver set the pattern for many other top jazz musicians in 1918 when he regularly "doubled," playing from 9:30 p.m. to 1:00 a.m. at the Dreamland Cafe before moving up to the Pekin Cafe at 27th and State from 2:00 to 6:00 a.m. Those whose instrumental virtuosity or technical skills helped them to land theater jobs "tripled," playing as many as five shows in a pit band, a cabaret show until 1:00 a.m., and winding-up with four hours in an after-hours club like the two Elites, the Pekin, and the Apex Club. Such schedules demanded far more than fleeting inspiration: endurance, discipline, and a deep commitment to music were required.[12]

Around whites in the black-and-tan cabarets, black musicians learned to be cooly aloof, disciplined craftsmen, adept at cordial restraint. Likewise, bandleaders disciplined spontaneous behavior of their sidemen. Joe Oliver took a brutally direct approach to inculcating northern discipline and regimentation—he placed a pistol on his music stand. As he wrote to one musician:

> This is a matter of business, I mean I wants you to be a band man, and a band man only, and do all you can for the welfair [sic] of the band in the line of playing your best at all times.[13]

But adaptations penetrated below the surface level of dress and show business strategies. Substantive musical issues were also involved. Musicians now faced more varied and lucrative performance opportunities requiring a variety of new musical skills, and, of course, a more conscious manipulation of older ones. The range of possibilities was remarkably wide and included enhanced instrumental performance techniques; a new, complex jazz and popular song literature; and South Side bandleaders' artful admixtures of aural improvisers with literate musicians.[14]

At one extreme, as many an African-American musician from W. C. Handy and James Reese Europe to trumpeter Lee Collins discovered, white audiences, nostalgic for an earlier, supposedly simpler, non-technological world, could be seduced with musical primitivism. Handy, "Father of the Blues," learned this lesson in the South when a crowd for whom he was performing requested "some of 'our native music'" played by three local blacks on a battered guitar, mandolin and a worn out bass. This trio played "one of those over-and-over strains that seem to have no very clear beginning and certainly no ending at all ... the kind of stuff ... associated with cane rows and levee camps." This "haunting" music brought a shower of silver dollars from the dancers and convinced Handy that there was money in folk music. "That night a composer was born, an American composer." Orchestra leader James Reese Europe, in order to maintain the illusion of the "naturally gifted" black musician, would rehearse his band on stock arrangements, leave the scores behind, and, when taking requests for these thoroughly rehearsed tunes, ask customers to whistle a few bars, and then "confer" with the musicians "in order to work it out with the boys." In 1931 at the Paradise Club on North Clark Street, trumpeter Lee Collins marveled at the popularity of Joe Stacks' skiffle band of folk "drifters" brought north by a rich New Yorker; Stacks flashed a roll of bills "big enough to choke an alligator."[15]

In Chicago, any genuine musical primitives would have lacked the instrumental skills necessary to adapt to varied playing situations; more versatile jazz performers learned to include primitivist acts in their repertoires. Skilled musical craftsmen like drummers Jimmy Bertrand, Warren "Baby" Dodds, and Jasper Taylor easily accommodated themselves to these minstrel and vaudeville traditions by placing thimbles on their fingers and fabricating rachety, shuffling sounds on washboards for novelty recordings like those of Jimmy Bertrand's Washboard Wizards. This same element of calculated primitivism suffused the entire northern approach to "jazz," but Chicago jazz performers refused to be typed as crude barroom or street corner entertainers. Joe Oliver's advertisement of his jazz band as "Eight Men Playing Fifteen Instruments" expressed some of the musical ambitions of South Side jazz.[16]

Ambitious jazz musicians, for example, took new pride in performing on elegant, up-to-date instruments. Arriving musicians soon discovered that representatives of the musical instrument manufacturers were willing to lend (or sell at a discount) new instruments in exchange for testimonials. Well-publicized leaders could profit most from this arrangement, and proven instrumentalists, who appeared in publicity photos wearing tuxedoes, also received new musical instruments with the understanding that they place them in the foreground. In fact, the musical instrument companies sometimes paid for half the cost of such publicity with the understanding that their brand name would be used on record labels and cabaret announcements. A photo of Cook's Dreamland

Orchestra, part of an advertisement in the *Defender* for the Tom Brown Music Corporation in the State-Lake Building, appeared over the band's endorsement of Buescher Musical Instruments. Like the other cabaret stars, musicians prized the new, highly polished, gleaming instruments. Louis Armstrong recalled that, before moving to Chicago, he, like most New Orleans jazzmen, thought "you had to be a music conservatory man or some kind of a big muckity-muck to play the trumpet." He got over such folk myths in Chicago, where he switched from the cornet to the trumpet.[17]

But Chicago's transforming power went beyond the mere possession of a new musical instrument. Cabaret musicians had to be accomplished enough on these "legitimate" musical instruments to be able to "play the show," both as a featured act and in accompaniment to vocalists, dancers, and comedians. New Orleans musicians who couldn't discipline themselves enough to do this didn't make it in Chicago. Bill Bottoms, who at first refused to book anything as raucous as a jazzband, had to fire trumpeter Tig Chambers when his band proved incapable of playing the Dreamland cafe's shows.[18] Lesser New Orleans musicians who had come north on someone else's coattails faded quickly from the scene. Technically limited musicians could easily play a short feature in the show, but could not accompany more sophisticated vocal acts, which depended heavily on the manipulation of tonal colors, nor could they generate the range of moods required for dance features.

Some cabaret floor shows produced for white audiences could become elaborate revues of original songs, dances, musical features all tied together by a theme which drew upon images of southern plantations, Mississippi River levees, or paddle wheeling show boats. In 1922, white promoters Morris Greenwald and Jimmy O'Neill mounted an elaborate African-American show called "Plantation Days" at Chicago's Green Mill Gardens on the North Side. This forty-five-minute revue had a cast of twelve principal actors, a chorus line of six dancers, and Charles Elgar's orchestra of twenty-six pieces. Playing for a show like this required the ability to read scores designed to accompany a variety of on-stage scenes. When Elgar wouldn't leave Chicago, "Plantation Days" went on tour to London with an orchestra led by New York pianist/composer James P. Johnson.[19]

Similarly, the Sunset Cafe built a complex revue called "Rhapsody in Black" around George Gershwin's "Rhapsody in Blue," and the Grand Terrace hired Percy Venable to design even more complex shows than he

had created for the Royal Gardens. Earl Hines testified that the elaborate
planning of calculated effects for the shows he played "helped me get
wrapped up in music, and they made me feel there was a big future in
it."[20] A handbill from the Plantation Cafe in 1926 gives an idea of the
range of musical roles—from accompaniment of vocalists to accompaniment of dancing chorus lines and roller skaters, to minstrel parading,
social dance music, and performance features—required by some of the
more elaborate cabaret floorshows:

Edward Fox, Mgr., Presents
Plantation Cafe's 3rd Anniversary Celebration, November 9,
1926
The Hottest Show in Any Chicago Cafe
Produced by Norman Thomas
Featuring the Season's Current Hit "Minstrel Days,"
A Real Old-Time Minstrel First Part.
Watch for the Big Parade, "Dr. Jazz,"
Based on Joe Oliver's Latest Song.
Spice—Pep—Girls Galore
New Principles—New Girls—New Comedians
Also see the Novelty of the Season, Bamboo McCarver,
Champion Roller Skating Dancer of America
Some of the Stars that Will Shine:
Naomi Thomas—Walter Richardson—Lulu Belle
Frankie Jaxon—Fred Andrews—Jackson and Bell—Red
Simmons—St. Claire Dotson
8 Plantation Ginger Snaps
Dance to the Entrancing Strains of Joe Oliver's
Dixie Syncopators[21]

Most South Side cabarets usually produced somewhat less elaborate
floorshows: the black cabaret entrepreneurs generally had less money to
invest in entertainment than did whites. The limited entertainment dollar of the ghetto public also encouraged the extinction of the "big revue"
on the South Side. *Variety* reported that "the colored amusement patron"
refused to commit to long-running shows of any kind, preferring "constant variety in the styles of his play fare." White promoter Robert Levy
experimented with many different kinds of shows but discovered that
none of them would work for more than a short run: "Levy says the
colored show patron is the craftiest shopper since Noah's wife went out to
buy her rainproof bathrobe."[22]

Moreover, many South Siders had serious reservations about the

"Down South," "Plantation" themes of elaborate "colored" shows like "Plantation Days." Such productions were so firmly anchored in racial clichés that black performers found it difficult to focus attention on their performance art. When, for example, the Entertainers Cafe re-opened after its losing battle with urban reformers and prohibition agents, the *Defender* objected to the title of its new floorshow "Plantation Review," which seemed to suggest "'down home' routines ... the title may seem a bit far-fetched to some who associate river boats and cotton bales with everything that hints of Dixie." For a number of reasons, therefore, cabaret managers preferred to feature a few separate, frequently changing vaudeville acts, cut out the expensive, specialized chorus line dancers, and use the musicians as both a stage feature and dance band.[23]

Playing a less pretentious floorshow could help musicians, vocalists, and dancers focus audience attention on their individual and collective performance skills. Musicians had to follow the cues that structured the sequence of events on stage; they had to remember agreed upon tempi and key signatures for vocal and dance numbers; rehearsals were required to perfect the timing of a wide range of special instrumental effects for comedy routines; accompanists coordinated carefully with tap dancers' rhythmic "catches" and jumps; and musicians generally had to learn to enhance other performers' acts with supportive, but unobtrusive, music. None of this necessarily required that every musician be musically literate, but adjusting to these playing situations often did mean that at least one musician, usually the pianist or the leader, sight-read music, since touring vaudeville and cabaret artists brought their own scores and arrangements with them. Non-reading sidemen had to learn the routines by ear and memorize them, thereby sharpening their aural analytical skills.

Chicago jazz performers also had to develop greater instrumental agility in order to play at much faster tempi than southern musicians. Late at night, with the lights turned down, black and black-and-tan cabaret musicians might play a grindingly slow New Orleans blues or a lament in a minor key; but show time audiences expected hot, fast musical action full of "pep" and "ginger," and in touch with the agitated rhythms of urban life. As banjoist Johnny St. Cyr, who was in a position to know, put it: "... the Chicago bands played only fast tempo ... the fastest numbers played by old New Orleans bands were slower than ... the Chicago tempo." Earl Hines agreed that, "we certainly played more up tempos in those [Chicago] days ..."[24]

In addition, those who performed in South Side cabarets often accompanied female vocalists like Alberta Hunter and Ethel Waters, cabaret stars who sang the latest in popular songs. These vocalists challenged their musicians with a new, rapidly changing repertoire of popular song material that was pouring forth from northern publishing companies. Popular songs created a major challenge to Chicago jazz performers; if they couldn't read the sheet music, and many could not, they had to be "fast," both in analytical grasp and instrumental technique. King Oliver, for example, had trouble playing the shows, but got by on his exceptionally sharp memory. As Jelly Roll Morton said of Oliver:

> My God what a memory that man had. I used to play a piano
> chorus, something like *King Porter* [Stomp] or *Tomcat* [Blues], and
> Oliver would take the thing and remember every note. You can't
> find men like that today.[25]

Louis Armstrong testified that Bessie Smith also possessed a remarkable memory, since "she'd always have the words and tune in her head."[26] Musicians like these were used to remembering tunes by playing them. Repetition engendered a functional recall of melodic, harmonic and rhythmic patters, even when, on a given occasion, mention of the tune's title brought nothing whatever to mind. Once one had begun to play, the tune flashed back into memory. Lil Hardin described how this mental process worked in a story she liked to tell about her audition with the Creole Band:

> When I sat down to play I asked for the music and were they surprised! They politely told me they didn't have any music and furthermore never used any. I then asked what key would the first number be in. I must have been speaking another language because the leader said, "When you hear two knocks, just start playing."
> It all seemed very strange to me, but I got all set, and when I heard those two knocks I hit the piano so loud and hard they all turned around to look at me. It took only a second for me to feel what they were playing and I was off. The New Orleans Creole Jazz Band hired me, and I never got back to the music store—never got back to Fisk University.[27]

South Side jazzbands of the late teens and early 1920s used a limited number of chords and usually played in the flatted keys, so it would have been possible for a pianist to guess at both the key and the probable harmonic progressions and quickly adapt.

But in Chicago, such casual guesswork increasingly gave way to more calculated approaches to music making. A musician's memory was often severely taxed by the unprecedented number of new popular songs issuing from music publishing houses and record companies. Although scarcely unknown in the New Orleans repertoire, the standard thirty-two-bar popular song was now as important to cabaret floorshows and dance halls as the New Orleans marches, which sounded less relevant inside a cabaret than in church or out in the street. According to *Variety,* show business success demanded "the creation, in a commercial sense, of a constant flow of new tunes, new musical ideas and novelties." Chicago Jazz quickly developed its own distinctive, varied musical literature, an amalgam of New Orleans tunes with newer original materials and popular songs. Musicians had to memorize the key signatures, melodic and harmonic patterns, and most effective tempi of a rapidly growing literature sung by a changing group of vocalists, which included Mamie, Clara, and Bessie Smith, Victoria Spivey, Ida Cox, Ma Rainey, Ethel Waters, Alberta Hunter, Lucille Hegamin, Lillie Delk Christian, Mae Alix, Mary Straine, Mary Stafford, Mattie Hite, and Josephine Stevens, in addition to Ethel Waters and Alberta Hunter.[28]

Many of the new tunes that seemed to lend themselves most naturally to jazz interpretation came from the pens of black songwriters who migrated to the South Side; their melodies, which formed the core literature of Chicago jazz, reflected the strong entrepreneurial spirit of the black ghetto in the 1920s. The earliest pioneer Chicago musicians built hopes on the inspirational success of pianist-composer-publisher and real estate investor Joe Jordan, who arrived in Chicago in 1903 from St. Louis, where he had learned ragtime at Tom Turpin's Rosebud Cafe. Musicians like Jordan and pianist Dave Peyton were the most musically literate of Chicago's pre-World War I black cabaret musicians. They were craftsmen who often were hired to compose or transcribe original materials for less literate performers. Peyton even advertised "will write orchestrations, songs taken from voice ..."[29] Jordan became musical director of Robert T. Motts Pekin Temple of Music from 1903 to 1912. He organized the Pekin Publishing Company to provide a publication outlet and copyright protection for black talent. He published several of his own works—"Pekin Rag" (1904), "J.J.J. Rag" (1905), "Oh Liza Lady" (1908), and "Dixie Land" (1908)—and broke ground in 1916 for the Jordan

Building, a three-story combination retail store and apartment building on the northeast corner of 36th and State streets. In 1917, Jordan successfully sued the Original Dixieland Jazz Band for appropriating his own "The Teasin' Rag" (1909) as part of their "Original Dixie Jazz Band One Step," frequently mentioned as the first jazz record.[30]

The leading postwar musician-entrepreneurs like Clarence Williams, Spencer Williams, and Jelly Roll Morton provided an original, challenging repertoire for Chicago's Jazz Age by recording, writing, and publishing compositions that combined blues with ragtime and popular song forms, converting more whimsically unorganized oral traditions into the uniform, standardized system of published European notation. In the process, Chicago jazz entrepreneurs transformed the country blues' metrical variety of thirteen-and-one-half or fourteen-and-a-half-bar motives, fragmented lyrics, and changing melodies into more appropriately uniform products. Blues historian David Evans argues that urban musicians who had learned European notation—either blacks themselves or those with access to cooperative whites who could write music—standardized the blues into the twelve-bar pattern, providing thematic lyrics that told a relatively coherent story, and melodic content focused into a fixed verse-chorus pattern of Tin Pan Alley popular songs. According to race record producer Mayo Williams, African-American pianists Lovie Austin, Tiny Parham, and Thomas A. Dorsey worked with performers in the recording studios, learning their original songs by ear and then writing them down in European notation.[31]

In 1916, William Christopher Handy's "Memphis Blues" (actually a combination of rag and blues strains) and "Jogo Blues" were on sale at the Frank B. Jones Music Company at 3409½ South State Street. New Orleans song writer, publisher, entertainer, and vaudevillian Clarence Williams and his partner, composer Armand J. Piron, moved to Chicago in 1918 and opened the Williams and Piron Music Company ("Home of Jazz") at 3129 South State Street. Soon they were selling sheet music for "Uncle Sam Ain't No Woman, but He Sure Can Take Your Man" (A Big Jazz Blues), "Ragtime Dixie Ball" (Great Jazz Hit), and "No More Cabarets in Town." Williams left for New York after the Race Riot of 1919, but his music store stayed open; he returned to Chicago frequently, especially to plug new songs like "Royal Garden Blues" (1919). As he put it:

> The demand for jazz music never was so great as it is right now.
> All our numbers are doing well, but we predict a record sale for

our three new numbers, judging from the avalanche of orders received already. The people want good jazz music and that is what we aim to give all the time.[32]

Joe Oliver also involved himself in this form of musical commerce by marketing the southern musical tradition in the North. As he wrote to fellow cornetist Buddy Petit:

If you've got a real good blues, have someone to write it just as you play them and send them to me, we can make some jack on them. Now, have the blues wrote down just as you can play them, it's the originality that counts.[33]

Jelly Roll Morton's many original compositions published in Chicago by the Melrose Brothers clearly indicate the entrepreneurial and commercial dimensions of jazz in Chicago. When Joe Oliver introduced Morton's "Wolverine Blues" in the spring of 1923, he received many requests for copies of the published sheet music. Walter Melrose, then running a "little old dirty [music] shop" and looking for ways to make money in music, discovered that the number had not been published. Oliver told him that Morton was living on the West Coast, and Melrose wired the composer a sizable advance in exchange for permission to publish "Wolverine Blues." Walter, Lester, and Frank Melrose, white musical entrepreneurs who were also said to act as go-betweens between black musicans and the record companies, published Morton's most memorable numbers—"Mr. Jelly Lord," "The Pearls," "London Blues," and many more.[34]

Chicago's resident jazz composers used cabaret floor shows and vaudeville as vehicles for the introduction and promotion of their latest songs. According to one observer,

[Clarence] Williams kept on plugging until he could land his songs in different acts and until quite a few were singing his songs. This caused the publisher to take notice and the Shapiro-Bernstein Music Company took over a batch of Williams' songs that were big hits later.[35]

Orchestra leader Erskine Tate plugged Clarence and Spencer William's "Royal Garden Blues" at the 1500-seat Vendome Theater by performing an arranged version in which each section of the orchestra, hidden in different parts of the theater, began "Royal Garden Blues" when cued by the "old familiar minstrel roll off," and played it while marching toward the pit.[36]

The club by the same name also featured the number and promoted other "hits" to issue from the Williams and Piron Music Company. For example, the Royal Gardens promoted May 16, 1919, as "'Jelly Roll Night!' featuring Clarence Williams' 'I Ain't Gonna Give Nobody None O' This Jelly Roll,' a free piece of jelly roll with every order, and the Music of the World's Greatest Jazz Band." Similarly, the club held "New York Night with the Lafayette Players" to honor the all-black theatrical repertory troupe, which had brought legitimate theater to the South Side. The "Famous New Orleans Jazz Band" featured "Their Great Patriotic Number" on this occasion, and Clarence E. Muse presented a demonstration of make-up techniques.[37] Thereafter, the Royal Gardens presented vaudeville acts on a "specially built stage"; jazzbands played for dancing between shows. In 1923, the Sunset Cafe adopted a racetrack theme, dressing the waiters in jockey uniforms, and featured book and lyrics by Clarence Muse and music written by Joe Jordan and played by Carroll Dickerson's Orchestra.[38]

Professional jazz performers also studied instrumental expression, tinkering with techniques in order to develop a performance specialty, a distinctive sound or instrumental "act" that would make what cabaret performers called an "up" (vaudevillians spoke of a "turn") during floor-shows. A featured entertainer earned between $15 and $35 a night in the early twenties, substantially more than sidemen in the orchestras. Some musicians mixed comic vaudeville effects with instrumental performance in their acts. Drummer Baby Dodds shimmied his stomach muscles in time to the tight press rolls he played on his deep-voiced trap drum. Bassist Bill Johnson lay down and played his string bass on his back or while lying on his side. Even the symphonically inclined orchestra leader Erskine Tate encouraged musicians like bassist Milt Hinton to "... lay down with it—he used to say, 'lay down on the [stage] floor.'" Other instrumental effects were less visually outrageous: Johnny Dunn, whose family lived in Chicago while he worked widely in Europe and America, billed himself as "Originator of 'Trumpet Tricks,'" and specialized in the "wah-wah," mute, and an elongated, four-foot "coach trumpet" which he had discovered in London. Sidney Bechet used his lips on the mouthpiece and reed to get "effects like chicken cackles." Joe Oliver also featured vocal effects, but used the cup mute. Clarinetist Jimmy O'Bryant recorded "hotsy-totsy" music, cajoling a satiric, snickering laugh from his instrument. Earl Hines performed requested tunes at a midget piano on coasters, moving about from table to table at the Apex Club.[39]

Such gimmicks reflected an adaptation of vaudeville comedy to the night club, but musicians often included revealing social commentary in their comedy routines: Oliver and Johnson developed an act for the Royal Gardens during which the cornetist worked his cup mute to produce a series of sounds like baby talk, and Bill Johnson, who appeared to be Caucasian, would sooth "the baby" in his high-pitched voice. As Louis Armstrong remembered:

> That first baby was suppose[d] to be a white baby. When Joe's horn cried like the white baby, Bill Johnson would come back with, "Don't Cry Little Baby." The last baby was suppose[d] to be a little colored baby, then they would break it up. Joe would yell, Baaah! baaaaaaah! then Bill would shout, "Shut up you 'lil so and sooooooooooo. Then the whole house would thunder with laughs and applauses.[40]

Instrumental imitations of vocal inflections can be interpreted as adaptations to northern, urban mechanization by "... assimilating the voice into the realm of the instrumental: to make it, as it were, all appendage to the machine."[41] While some northern admirers of black southern music often interpreted the voice-like sounds produced by jazz performers as rejections of the original designs of European musical instruments, such sounds can be seen just as readily as resulting from a continuing adjustment of oral communication to particular musical instruments.

For example, South Side musicians worked to interiorize technology, to make their chosen musical "machine a second nature, a psychological part of himself or herself." While their unconventional approaches to musical instruments have received much commentary, theirs was an encounter with musical technology. Jazz musicians learned how to make these tools do what they could be made to do and shaped themselves to these possibilities.[42] This process was not perceived as a dehumanizing capitulation to machine age regimentation but rather another enrichment of black traditions in music. In fact, the ultimate goal of ambitious jazz performers was a seemingly effortless, personal expressiveness in instrumental control. In this process, the top Chicago jazzmen of the twenties came to terms with various dimensions of European instrumental traditions.

The continuing encounter of musicians with their instruments produced intriguingly varied results: some jazz instrumentalists like Joe
Oliver, Lee Collins, and Johnny Dodds arrived at intensely personal,
unusual, individualistic styles that were relatively less dependent upon
traditionally defined instrumental technique than the jazz styles of Louis
Armstrong, Jimmy Blythe, Earl Hines, Jimmie Noone, and Jimmy
Bertrand, whose playing came to be distinguished by a marked refinement of touch and tone. Mixing musicians of different stylistic tendencies
gave Chicago style jazz much of its distinctively varied flavor.

South Side Chicago jazz performers also differed widely in their
comprehension of the fundamentals of musical notation and traditional
instrumental technique. Cabaret and dance music showed two interrelated, but distinct, lines of development: formal musical education and
more informal jazz apprenticeships. Some cabaret musicians, such as
clarinetists Jimmie Noone and Buster Bailey, studied formally with
Franz Schoepp, the white Chicago clarinetist who also taught Benny
Goodman. Reedmen Jerome Pasquall, Darnell Howard, Omer Simeon,
and Clifford King similarly pursued formal instruction both within and
outside of the South Side ghetto. Dance band leader Charles "Doc"
Cooke, in whose orchestra Jimmie Noone was a mainstay, earned his
doctorate in music from the Chicago College of Music, and presented a
copy of his dissertation composition "Pro Arte" to William Abbott.[43]

Major N. Clark Smith, at one time musical director at Tuskegee
Institute, headed the music program at Wendell Phillips High School on
the South Side. Smith, a strict disciplinarian who "didn't dig it when you
were playing jazz," organized a youth band for the Chicago *Defender* and
several second-generation Chicago jazz performers—bassist Milt Hinton, percussionist Lionel Hampton, bassist Hayes Alvis, who worked
with Jelly Roll Morton and with Jimmie Noone—took his high school
courses. Smith had once worked for the Lyon and Healey Music Company, which subsequently donated musical instruments to the South Side
high school.[44]

Many jazz performers joined the 370th Division, Eighth Illinois In-
fantry National Guard Band in order to improve their skills, or, as
reedman Albert "Happy" Caldwell put it, "to get some good, you know,
learning, playing all that type of music." Moreover, many highly trained
musicians like Erskine Tate, Charles Elgar, Dave Peyton, and Charles
Cooke supplemented their earnings by giving private lessons, so that
many jazz musicians assimilated theory and instrumental technique dur-

ing the day, while also playing far into the night. Columnist Dave Peyton, outspoken advocate of the professionalization of music, proposed the licensing of all private music teachers, claiming that many ambitious musicians were learning unorthodox techniques.[45] Peyton would have formalized all musical instruction, but his organized professionalism never erased the continuing traditions of apprenticeship learning which encouraged many jazz musicians to preserve and extend African-American playing styles.

Some improvisers learned about music from the more formally trained musicians in the bands with which they worked. Louis Armstrong survived the Streckfus brothers' floating conservatory thanks to tutoring from fellow bandsman, mellophonist David Jones. Later, he further benefited from hymn-reading sessions with Lil Hardin, when they worked together in the Oliver band. They met and later married in Chicago.

Pianists like Hardin played a particularly important role in infusing immigrant New Orleans dance music with new insights taken from the largely forbidden secrets of legitimate music. Pianists Tony Jackson, Jelly Roll Morton, and Richard M. Jones all came to Chicago from New Orleans, but a majority of the more prominent South Side pianists—Lil Hardin, Lovie Austin, Earl Hines, Alex Hill, Teddy Weatherford, Zinky Cohn, Jimmy Blythe, Luis Russell, Tiny Parham, and Casino Simpson— did not hail from the Crescent City. They arrived in Chicago from a remarkably broad range of geographical locations: Memphis, Chattanooga, Pittsburgh, Little Rock, Bluefield, West Virginia, Oakland, California, Louisville, Kentucky, Panama, Winnipeg, Manitoba, and Chicago respectively.[46]

South Side jazz instrumentalists of the 1920s learned a wide repertoire of jazz techniques in informal apprenticeships to older jazz performers. Louis Armstrong's tutelage with the man he called "Papa Joe" Oliver is well known. Clarinetist and tenor saxophonist Happy Caldwell listened carefully to Buster Bailey's playing in the Carroll Dickerson band and spoke of a "school" of clarinetists—Bailey, Jimmie Noone, Omer Simeon, Cecil Irwin, Ed Beckstrom, and Darnell Howard—who hung around together and influenced one another's work. Bassist Bill Johnson taught Milt Hinton how to "slap" the bass so that the strings snapped against the instrument's neck. Hinton then developed his own specialty: "... instead of just using one slap [per beat] like he used, I would multiple slap it—triples and quadruples." Drummer Jimmy

Bertrand, who had studied drums at the Catholic school on 55th and Halstead streets and with "old man Johnson," father of Erskine Tate, as well as Roy Knapp of the Minneapolis Symphony, taught jazz drummers Lionel Hampton and Sidney "Big Sid" Catlett.[47]

New show business tricks, instrumental dexterity, popular song literature, and harmonic awareness did not necessarily indicate, as one scholar concluded, that the South Side musician was "suppressing the traits of his own subculture and acquiring the traits of the dominant white middle class."[48] New techniques and ideas were blended into African-American musical culture in original and swiftly changing ways. As a historian of popular culture in early modern Europe has put it: the "minds of ordinary people are not like blank paper, but stocked with ideas and images ... Traditional ways of perceiving and thinking form a kind of sieve which will allow some novelties through, but not others."

Even when studying music formally the Major Smith, Franz Schoepp, or at the Chicago College of Music, ambitious musicians adapted European techniques and ideas to their own uses. The encounter between elite and popular musical cultures produced an alloy called "jazz," proof that music students were rarely passive consumers.[49] South Side musicians made individual adjustments to white musical culture, and their levels of assimilation reflected a variety of different factors: how old they were upon leaving the South; how long they stayed in Chicago or another big northern city; the sort of contacts they had or could hope to have in the profession; and other more personal and aesthetic factors.

The often underrated Chicago clarinetist Darnell Howard played a vital role in the growth of technical sophistication and versatility in South Side jazz. Born in Chicago to musical parents around 1895, Howard took up the violin at age seven and began his formal musical education under Charles Elgar. The dean of black Chicago orchestra leaders, Elgar organized many classical music concerts on the South Side, gave music lessons, and led a dance band at Harmon's Dreamland dance hall from 1916 to 1922. Howard performed from an early age with Elgar's student orchestra, with whom he was featured during a May 21, 1912, performance of Donezetti's *Lucia di Lammermoor.*

After working extensively in dance bands, with smaller formations at the Loop's Lamb's Cafe and the South Side's Elite Cafe, and touring the Midwest with Elgar, Howard joined the "Plantation Days" pit band with several other Elgar sidemen for a trip to London in the spring of 1923. On his return, Howard played in Carroll Dickerson's dance band and Dave

Peyton's Plantation Cafe Symphonic Syncopators, and made the transition to King Oliver's Dixie Syncopators with whom he played alto and soprano saxophones, clarinet, and violin.[50]

Howard bridged the worlds of concert hall and cabaret music, the first of a larger group of versatile instrumentalists that included reedmen Jimmie Noone, Buster Bailey, Omer Simeon, Barney Bigard, and Cecil Irwin, tubists Hayes Alvis and Quinn Wilson, trumpeters Shirley Clay, George Mitchell, and Homer Hobson, pianists Hines, Cohn, Hill, Weatherford, and many others, musicians who made the arranged elements of Chicago jazz possible while playing in the shadow of the solo stars.

As Darnell Howard's career indicates, the South Side movie theater and dance hall orchestras brought increased technical sophistication to those who played in them. Thomas J. Hennessey has delineated a musical field, closely related to Chicago's cabarets, occupied by Charles Elgar's Creole Orchestra, Erskine Tate's Vendome Theater Orchestra, Charles Cook's Dreamland Orchestra, Sammy Stewart's orchestras, and Clarence M. Jones' Owl Theater Orchestra. It emphasized the legitimate, European theoretical and instrumental approaches to music which these large aggregations brought to Chicago. These bands "... co-opted, at least locally, the ground which elsewhere the new jazz innovators would develop into the swing style."[51]

But it is misleading to present such groups as a musical "establishment" in "conflict" with the jazzmen. The tendency to see large bands playing arrangements as somehow in conflict with "the New Orleans men" in Chicago stems from an inability to see the varieties of musical approaches and skills possessed by Chicago musicians and the creative ways in which they blended them. The South Side arrangement-reading big bands represented the possibilities in a "legitimate" polarity at one end of a continuum of musical skills and philosophies that extended through several of the more innovative jazz groups that made records in the late twenties, to the polyphonic improvisations of jazz band musicians.

Of course, the theater and large dance hall orchestras did inhabit a musical world far removed from the novelty washboard bands. As Hennessey and Hsio Wen Shih have argued, they represented a more middle-class respect for the refined technique, musicianship, and concert hall repertoire associated with the surrounding white culture. A new breed of college-trained African-Americans, born in border states of the South, mixed harmonic sophistication, popular song form, and stock and original arrangements with field hollers, spirituals, and the blues. But African-American traditions in music were quick to absorb, rather than reject or segregate new influences. As reedman Garvin Bushell told music historian Mark Tucker, Chicago cabaret bands were not limited to improvisation:

> Chicago jazzmen had the advantage in those days of having a crack at theater music before the New York jazzmen did. They improved their ability that way, and so could read a little better than jazz musicians in the East.[52]

Some immigrant musicians—indeed, some of the most famous, made only partial adjustments to musical literacy. Louis Armstrong apparently learned to read music while an apprentice on the Mississippi paddle wheelers. But many of his colleagues—bassist Pops Foster, drummer Baby Dodds, trumpeter Lee Collins, reedman Sidney Bechet, and many others—learned musical notation only incompletely, sometimes insisting that they were able to "spell," if not fully read, music. For some, the musical alphabet acted as "a major bridge between oral and literate mnemonics," but others found it a barrier. In some senses, these musicians didn't have to adjust: each improviser contributed such a powerfully individual voice that bandleaders could build around them.[53]

The struggles between Dave Peyton, the composer, Chicago *Defender* columnist, and theater orchestra leader, and the jazzband leader Joseph Oliver, illustrated the key issues of cross-cultural synthesis at stake in Chicago during the twenties. The outspoken advocacy of European concert music found in Peyton's weekly column marked the farthest extreme in the assimilationist interpretation of South Side music. Peyton, who directed the pit band at the Grand Theater, advised all South Side musicians to abandon "gut bucket" cabaret music with its "squeaks, squawks, moans, groans, and flutters"; Peyton insisted that a "jazz-crazed public" and the "hip liquor toter" had created the demand for such "novelty," "hokum" music, leading race musicians to abandon concert hall instruments like the violin for louder, more vulgar instruments like the banjo. The day was coming, he felt sure, when race orchestras would break into the vaudeville and movie theaters throughout the city as regular contracted pit bands; he called upon all musicians to prepare themselves by studying theory, harmony, and proper instrumental technique.[54]

The relentless, often brittle edge to Peyton's arguments led him into repeated attacks on cabaret orchestras in general. While praising Erskine Tate, Clarence M. Jones, Lovie Austin, and the other theater band leaders, he condemned most jazz playing as rank musical ignorance, insisting that any trained musician could do it, while jazz performers simply could not play legitimate orchestral scores. As the public seemed to demand it, Peyton agreed that theater stage and pit bands should include at least a little jazz in their performances, in the knowledge that the entire fad would soon pass.

Peyton's "The Musical Bunch" column in the Chicago *Defender* showed a recurrent interest in Joe Oliver's jazz bands, acknowledging them to be the best in a line of business which the writer scorned. By 1925, Peyton and Oliver competed for important South Side performance jobs: Peyton hoped to use an occasional cabaret engagement, like the one he secured for his own band at the Plantation Cafe, to educate cabaret customers to traditional orchestral music; while Oliver, and his protégé Louis Armstrong, pursued success within the world of cabaret music. Oliver, however, was by no means the musical hacker that Peyton associated with jazz, and the columnist knew it.

Oliver's rivalry with Peyton came to a climax in November 1925 when the latter hired him as a featured cornet soloist with his ten-man Symphonic Syncopators at the Plantation Cafe. Most accounts stress bandleader Peyton's reprimand of Oliver for attempting at one point to leave the bandstand ("'Look out, Joe, my musicians don't leave the stand without permission. You are in my band now, and must do as I say' ... 'I beg your pardon, Fess ... etc.'"), but more was at stake. Typically, Peyton had combined orchestral and concert hall music with jazz, and *Variety* reported that even those Plantation Cafe customers who were not interested in dancing "applauded the merits of the orchestra spontaneously. Their rendition of heavy operatic numbers is handled as easily as the ordinary syncopated tune."[55]

Early historians of jazz have emphasized that Oliver joined Peyton only after losing his own job in the December 1924 fire which destroyed the Lincoln Gardens, but Oliver had already joined Peyton's Symphonic Syncopators in November, a month earlier. Peyton, known for his long service in the pit at the Grand Theater, for booking bands, and for his newspaper column, was not a cabaret musician and needed the cornet star in his Plantation Symphonic Syncopators in order to appeal to dancers and jazz fans. As jazz cornetist Lee Collins discovered when he went to work for him, Peyton's band was a "stage band, not a real jazz orchestra. They played a lot of overtures and things like that and didn't swing."

Oliver soon took over the Symphonic Syncopators, renaming them the Dixie Syncopators, and played from 1925 to 1927 at the Plantation Cafe, replacing Peyton at the piano with Luis Russell and stealing musicians from Peyton's band from time to time. Oliver, therefore, epitomized the cultural amalgam that produced Chicago jazz: some of his band's featured numbers ("Too Bad," "Deep Henderson") were built on complex arrangements, some of his musicians (trumpeter Bob Shoffner, reed man Darnell Howard, and pianist Luis Russell) were perfectly at home reading scores, but Oliver mixed such "legitimate" musicians with those who could swing and solo regardless of their other skills.[56]

Louis Armstrong's invention of dramatic improvised instrumental solo statements marked the single, most outstanding contribution to the ongoing synthesis that produced jazz in Chicago during the twenties. Armstrong's wonderful creations surpassed contemporary comprehension. Even Peyton, the self-appointed musical sage who scorned jazz could only marvel:

> Louis Armstrong, the greatest jazz cornet player in the country, is drawing many Ofay [white] musicians to Dreamland nightly to hear him blast out those weird jazzy figures. This boy is in a class by himself.[57]

Armstrong's new musical synthesis combined three sources: African-American folk music traditions, elements of cabaret musical entertainment, and techniques borrowed from Anglo-American musical culture.

From oral folk traditions, he took his conviction that sounds have great power, especially when communicated in heavily rhythmic, balanced patterns of repetition and antithesis. Armstrong's solos, more than those of the white cornetist Leon "Bix" Beiderbecke, for example, possessed the eloquence and drama associated with the winners in those verbal struggles called "the dozens," where "signifying" contestants tongue-lashed opponents into silence. As what Ralph Ellison has called a "rowdy musical poet," Armstrong fabricated grandly improvised variations on traditional melodic, rhythmic, and harmonic patterns.[58] The principles of black oral rituals described by Henry Louis Gates, Jr., parallel some of his solo devices: musical metaphor (quoting a melodic pattern from one song during a solo variation on another); hyperbole (exaggerating a distinctive feature in a given melodic, harmonic, or rhythmic pattern); chiasmus (inverting the relationship between the elements in parallel phrases); and synecdoche (using a phrase element in place of the whole from which it was taken, or applying double time to fit a large part of a melody into a short fragment of time).[59]

In the cabarets, Armstrong learned what Chicago jazz historian John Steiner has called "the value of the instrumental soloist as entertainer."[60] Chicago show business demanded that musicians learn to make an instrumental "statement," something to show contractors, club managers, and band leaders. In this highly competitive environment, musicians developed jazz techniques very quickly. Solo statements inevitably grabbed the spotlight, acting as the ambitious musician's "act" and moment of glory. Jugglers briefly defied laws of time and space; vocalists crammed every trick they knew into one or two feature numbers; a comedian had three minutes to stimulate rolling waves of laughter; tap dancers dazzled the clients during their short features; and jazzmen, in turn, grew accomplished at transforming swiftly passing beats and harmonies with their own, personal inventions of timbre, tone, range, phrasing, and harmonic insight.

Lastly, Armstrong's remarkable solos depended upon an exceptionally high level of instrumental virtuosity, one that appeared more often in the concert hall. Armstrong developed an exceedingly wide instrumental range, impressive power, and technical dexterity with which to demonstrate sweeping, dramatic jazz solos of grand proportions. Although he never described how he did it, the solo star did mention practice sessions with Lil Hardin, who had studied music at Fisk University before coming to Chicago and continued on to earn her teaching certificate at the Chicago College of Music. Song writer Hoagy Carmichael claimed that Lil Hardin "got a book of the standard cornet solos and drilled him. He really worked, even taking lessons from a German down at Kimball Hall, who showed Louis all the European cornet clutches."[61]

Further evidence of Armstrong's studious approach to the artistic problems of the jazz solo emerges from two printed editions of his solo work, published by the Melrose brothers in Chicago. One of them, published as *125 Jazz Breaks for Cornet,* featured, according to the *Defender,* his particular solutions to "jazz endings" (also called "turnarounds," those concluding beats of a chorus that feature harmonic transitions back to the chorus beginning), "jazz connections" (moments of harmonic transition from the A to the B and back to the A sections in AABA popular songs), and jazz breaks. The other book offered *50 Hot Choruses for Cornet,* transcriptions of Armstrong's improvisations on songs sold by Melrose. As James Lincoln Collier notes, these books reveal that Armstrong's improvisations were primarily melodic and rhythmic rather than harmonically oriented.[62]

Virtuoso solo improvisation was bound to have been developed by someone, since adaptations of African-American music traditions to city culture pointed in that direction. But Louis Armstrong extensively demonstrated this breakthrough in twenties Chicago and imprinted it with his remarkable musicianship, doing something telling with every note he played, shaping music which was not merely flashily elegant, dramatic, outrageous, and hilarious, but structurally balanced, passionate, and beautiful. Armstrong's solo style illustrated an artful blend of African-American music with cabaret culture.

But the world of Louis Armstrong and the South Side black-and-tans functioned within a larger network of musical entertainment, a complex series of large racially exclusive dance halls and hotel ballrooms, where musicians played musical styles which, while resembling and even appropriating jazz in many ways, pursued more commercial directions suitable to a broader market. It is time to turn to this world of jazz age social dance music, which, even if considered as a sort of "near beer" by jazz lovers, passed for the real thing among thousands of Chicagoans.

3

White Jazz and Dance Halls

THE CORE CULTURAL and musical synthesis that produced Chicago jazz of the twenties emerged in the cabarets from 1904 to 1929. Impressive numbers of white Chicagoans, however, danced to Jazz Age music in the large Loop hotels and in the large dance halls built in the major North Side and West Side bright-light districts. Only one of these more sizable institutions—the Savoy Ballroom—was built in the black ghetto (and late in the decade); the large commercialized dance hall was overwhelmingly a white phenomenon that catered to a craze among the white population for social dancing. Dance halls and hotel ballrooms presented varied styles of dance music and entertainment against which jazz struggled to create a separate identity. For jazz musicians and their close listeners, and even for some dance band musicians, the "real thing" was always defined by both its resemblance to and differences from commercial dance music. Isham Jones, Chicago's premier dance band leader, whose orchestras recorded over 200 titles in the 1920s, refused to label his music "jazz," and claimed that most dance band musicians considered jazz a "'down South Negro type' of blues." In order to better advertise the broader range of sensibilities encouraged by his dance orchestra, Jones preferred that his music be called "American Dance Music."[1] The majority of Chicagoans probably were less sensitive to music, race, and urban cultures and would often accept jazz age dance hall music as jazz itself, particularly when a good dance band wanted them to.

Large commercialized dance halls of the Roaring Twenties catered to a much greater number of Chicagoans than the smaller cabarets, and presented several styles of social dance music other than jazz. They accentuated the physical activity of the customers' dancing, and avoided the cabarets' role as "speakeasies," where illicit alcoholic beverages were consumed. Even before prohibition, entrepreneurs of Chicago's largest dance halls organized to meet urban reformers' criticisms of small halls, which combined elements of the saloon, cabaret, and dance hall, places where social dancing mixed with the sale of alcoholic beverages, vaudeville-style entertainment, and sometimes prostitution. The large commercialized dance halls came to focus attention on morally restrained, often athletic social dancing and encouraged several brands of arranged, big band music designed to stimulate it.

From the start, jazz age music in Chicago was deeply intertwined with a mounting enthusiasm for social dancing that had swept through the rooming house neighborhoods. Young urban audiences of the second and third decades of this century were no longer content to sit passively and watch vaudeville musicians, vocalists, and dancers perform exciting new dance demonstration numbers on stage; they wanted to get closer to their musician-heroes, and even to play, sing, and dance to the new songs themselves. It is no surprise that jazz-related music caught on with the public in new leisure institutions: commercial dance halls in Chicago's neighborhoods made the customers into the stars of their own productions of energetic but urbane and sexually charged physical movements, which provided ample room for individual creativity within a shared public sensibility. The dance bands, their handsome leaders, and star sidemen made a varied popular music to which the public danced.[2]

White Chicago experienced a widespread, grassroots social dance movement which got under way by at least 1910. In 1911, for example, urban reformers began a series of worried public reports on dance hall evils. This popular phenomenon had roots in the late nineteenth-century concert saloons—those with an extra room for entertainment and dancing—where, during the ragtime era, solitary pianists and instrumental groups had set the city youth and the sporting set dancing to a transitional combination of ragtime with vaudeville comedy or novelty effects. In

1910–11, urban reformers had observed 328 small, dirty, ill-lighted, flammable, largely unregulated centers of entertainment, 190 of which opened directly onto a saloon. Liquor was sold outright in 240 others. Liquor was being sold to boys and girls fourteen to eighteen years old, who danced "tough dances" that often shocked their immigrant parents. Reformers fretted about the disreputable lodging houses located near the dance halls, where "innocents from Europe and the country are corrupted." In 1910, social dancing was more popular than going to the movies: as many as 86,000 youngsters a night thronged hundreds of obscure halls, most of which were thought to be controlled by "saloon and vice interests."3

Freiberg's on 22nd Street in the old Levee vice district had been "the most notorious of all places in the city where stepdancers gather," the prototype of the sort of night spot from which the leading ballroom and cabaret proprietors of the 1920s hoped to disassociate social dancing. Opened by Fritz Freiberg, the dance hall had catered to the "Bohemian" element or the "Sporting Set"—a mixture of politicians, vaudevillians, "thieves, whores, pimps, and gunmen, who mingled nightly with visiting business and professional men from out of town or respectable neighborhoods in the city itself." Ike Bloom, who managed Freiberg Hall and was later known as "King of the Brothels," had hired fifteen female dance instructors, who were ordered not to take their "tricks" out of the hall. Bloom later became an important Chicago night-life personality who, after the war, openly defied the wartime 1:00 a.m. Police Department closing order, and, when Freiberg Hall was closed, opened the highly successful Midnight Frolic where he featured "girlie entertainment" rather than music, defying the twenties trend toward clubs which featured music and dancing.4

Chicago's clubs during World War I often had been what newspaper columnist Westbrook Pegler called "dumps," small, poorly ventilated, dirty, makeshift cellars with close geographical and historical ties to Chicago's vice district. A leading reformer had this complaint about the lewd social dancing to the ragtime bands in clubs like these:

> Couples stand very close together, the girl with her hands around
> the man's neck, the man with both his arms around the girl or on
> her hips, their cheeks are pressed close together, their bodies touch
> each other; the liquor ... is like setting a match to a flame; they

throw aside all restraint and give themselves to unbridled licence
and indecency ... their animal spirits fanned to flame by the mad
music.

Similarly, in Chicago's cheaper neighborhood halls, the boys often
danced with their hats on, and smoked, drank, and even spat while idling
between numbers; the girls, who frequently placed powder puffs provoc-
atively in their stocking tops, sat on the laps of their beer drinking escorts,
and talked tough to fascinate the younger girls.[5]

The atmosphere of clubs and dance halls, filled with wild music, the
perspiration odors of the dancers, tobacco smoke, and alcohol fumes
fulfilled the worst nightmares of nineteenth-century moralists, who had
warned that social dancing would lead Christian youth directly to hell.
Nevertheless, the larger commercialized dance halls in the mid- to late-
teens and twenties responded to urban reform pressures on unregulated
night-life institutions, as well as to marketing decisions to appeal to a
broad cross section of society.

Mayor Carter Harrison, Jr.'s closing of the Levee vice district in 1912,
a response to pressure from the Committee of Fifteen and other reform
organizations, and continuing investigations by urban reformers into the
gambling, prostitution, and sale of alcohol at the earliest Chicago music
clubs of the years just before and during World War I, led to substantial
reforms and encouraged several sorts of more refined and commercialized dance music. The detailed reports of 1910 and 1917 authored by
Louise de Koven Bowen, Chicago heiress, socialite, philanthropist, Hull
House benefactor, and leading urban reformer, and her researcher Jessie
Binford, documented the behavior and the night-life institutions of the
formative years of Chicago's Jazz Age, when the "dance craze" encouraged a transformation of ragtime into jazz; her leadership of the movement to reform new night-life institutions, moreover, played an influential role in creating large specialized dance halls and peppy but refined
dance orchestras.[6]

As a founding member of the Juvenile Protective Association, created
in 1904 in association with the Juvenile Court Society to protect Chicago
youth from the debilitating effects of commercialized vice, Bowen spearheaded a Victorian backlash against new forms of urban entertainment
and helped to shape the foundations of Chicago's Jazz Age. She deeply
feared the effects of mixing social dancing with alcohol consumption in
unsupervised dance halls located near cheap hotels and rooming houses.

Bowen saw jazz as another pernicious ingredient in a combustible mixture of indecency and self-indulgence. To her, jazz was social dance music that somehow encouraged "indecent" dancing, movements that defied the prohibitions of Victorian dance etiquette against wriggling the shoulders, shaking the hips, and twisting the body. She opposed the "Dip" (during which the man bent his partner backward until her head touched the floor, at which moment the woman would kick up one leg "so that [her] privates were shown"). Couples should "stand far enough away from each other to allow free movement of the body in order to dance gracefully and comfortably."[7]

Although accused of all manner of immoral influences by some conservative social critics, certain styles of jazz and near-jazz social dance music actually contributed to Chicago's efforts to clean up even the worst dives by offering an alternative focus to commercialized sex. Music that suggested, in an ill-defined, diffuse way, forbidden worlds of sexual excitement and alcoholic abandon still was not, itself, an overtly sexual or illegal act. City government, responding to conflicting pressures from the reformers and night-life entrepreneurs, found music a lesser evil which could be structured and reformed more readily than prostitution or alcohol consumption.

Conflicting social pressures on social dance music found expression in municipal regulations. From 1916 to 1921, for example, the Chicago City Council had groped toward definitions of various different night-life institutions in order to respond to the contrary pressures from leisure time entrepreneurs and urban reformers. Music turned out to be one important medium for compromise between these two contrasting pressure groups. Eager to institute a system of revenue-producing, graduated licence fees which would be keyed to the size of the dance floor, the City Council tried to sift through a bewildering complexity of new leisure time activities.

The focus was placed on what Council members, under conflicting political pressures, considered common, but unsavory, commercial contacts between female entertainers—vocalists, dancers, and dance instructors—and male customers in dance halls, cabarets, and saloons. A reform ordinance of 1916, for example, required that places of amusement that offered musical entertainers and social dancing demark separate areas in which the entertainers performed and the customers danced. Where space was limited, the two groups were not to occupy the same area simultaneously in order to limit what was considered to be a core prob-

lem: commercialized sexual contacts between entertainers and customers.[8]

Neither at the time nor thereafter was there much complaint about
social dance music, itself, as a music-making process. Rather, this kind of
popular music became a problem when joined with prostitution, drunk-
enness, and lewd dancing. Some ragtime, particularly when combined
with traditional dance numbers, might even act as a calibrated, culturally
acceptable form of entertainment. As a result, a Chicago city ordinance of
1917, which aimed at cabaret reform, specifically excepted instrumental
music from the list of prohibited forms of entertainment. Council's legal
adviser noted that in exempting instrumental music this ordinance would
prohibit patriotic vocal performances while permitting "noisy ragtime,
the jangle of the jazz band and the tom-tom with which is associated the
sinuous oriental music," but still emphasized that:

> There is no time within the memory of any of us when instrumen-
> tal music was prohibited by statute or ordinance at places where the
> sale of liquor is permitted ... there was never any demand for sup-
> pressing such music until the entertainments in such places were
> broadened out beyond what people had been accustomed to.

Clearly, ragtime and jazz music in cabarets presented newer styles of
more traditional, acceptable forms of entertainment than did disguised
prostitution; from the point of view of the cabaret entrepreneur, the
advantage of a musical (rather than a "girlie") act lay in the fact that
customers sat and listened, or danced, and had no morally dangerous
interactions with the musicians. Jazz was intended to be exciting, animat-
ing, effervescent music, but it would retain a relative innocence throughout the twenties. Whatever raucous barrelhouse sounds they made in the
insalubrious surroundings in which they sometimes found work, musicians, at least while performing, were too occupied with playing their
instruments to indulge in vicious activities. In a world of prostitution,
gambling, and illegal alcohol, music and dancing were lesser evils.[9]

In Chicago's most celebrated early adaptations of jazz, the most
popular groups removed some of the moral onus from their music by
mixing in liberal amounts of vaudeville humor. Novelty bands, made up
of white southerners from New Orleans, cast African-American rhythms
and instrumental improvisations in a nervous, comic, often slapstick
mold. In 1915, Tom Brown's Ragtime Band, an all-white group, came to

Chicago from New Orleans to play a four month engagement at Lamb's Cafe at the intersection of North Clark and West Randolph streets. Although the word "jass" or "jazz" did not originate in Chicago (it has been traced back to a San Francisco journalist writing in 1913), it was used to denigrate Brown's Ragtime Band, when, in order to discredit the successful invaders from New Orleans, the leader of the Chicago band playing opposite them at Lamb's Cafe got the musicans' union to launch a smear campaign, spreading the word that Brown's music was "jass," an obscene term normally associated with other matters in Chicago's Levee vice district. Undismayed, Brown made the most of the situation and approved an advertisement for his group that read "Brown's Dixieland Jass Band, Direct from New Orleans, Best Dance Music in Chicago." They attracted substantial attention with their slapstick novelty numbers.[10]

By far the most important white group to play these "sardonic," "impudent," "slaphappy" sounds, which Neil Leonard has called "nut jazz," in Chicago in the waning nights of the gaslight, concert saloon era was "Stein's Band from Dixie." They were discovered in New Orleans by Chicago night-club entrepreneur Harry James, who booked them into Schiller's Cafe, 318 East 31st Street at South Calumet Avenue just on the southern edge of the Levee vice district and the northern edge of the fast emerging black belt.

In a shuffle of personnel, cornetist Nick LaRocca, trombonist Eddie Edwards, and clarinetist Alcide Nunez left Stein's band and reorganized themselves into a new group by adding New Orleans drummer Tony Sbarbaro and clarinetist Larry Shields. This group moved on June 2, 1916, to the Café de l'Abbée at 419 S. Wabash, and finally on July 6 to the Casino Gardens as the Original Dixieland Jazz Band. One year later, the ODJB's brand of frenetic "nut jazz" caught on at New York City's racially segregated Reisenweber Cafe and led to the band's celebrated R.C.A. Victor recordings of 1917, and international fame as the "inventors" of jazz.[11]

This "nut jazz" had a lasting impact on the social dance music offered as jazz by Chicago dance bands. From 1919 to 1925, for example, such prominent groups as that which Paul Biese led in vaudeville, ballrooms, and hotels performed an overtly comical form of syncopated dance music. Whether performing with his Trio, several of whose records were actually augmented by wind instruments, or by the Paul Biese College Inn Orchestra, Biese played tunes like "Bow Wow," "Chili

Bean," "Timbuctoo," "Happy Hottentot," and "Dardenella" with clearly delineated saxophone melody adorned by a broadly slapstick tailgate trombone, laughing muted trumpet, and squealing clarinet. "Dangerous Blues," a promising jazz title, was played for laughs, featuring a satiric "dasdardly villain" introduction to a sweetly sentimental ballad appropriately entitled "Sweet Love." Biese's "Happy Hottentot," "Timbuctoo," and "Chili Bean," referred snickeringly to the allure of young African and Latin American women (who were, of course, barred from white hotels and dance halls). The "Happy Hottentot," according to the lyrics, is always happy to: "Meet you, greet you, glad to eat you; Shake you, bake you, and shimmie shake you."[12]

Dance styles were similarly shaped to middle-class white taste. Customers in the cafes of the late ragtime/early jazz years were helped to "get onto" the latest dance steps by imitating stage dancers who performed for them. In the all-white clubs of Chicago, the most influential male dancer was Joe Frisco, who was known as a "jazz dancer" as early as 1914 and made his reputation as the "American Apache," dancing in clubs and on vaudeville stages to music performed by Bert Kelly's Jazz Band or Tom Brown's jazz band; he was photographed with the latter of these groups (Al Capone looks on). Frisco, as he was known, demonstrated the fox trot with Loretta McDermott before changing into "tough clothes of the Bowery," wearing a smartly cocked derby, smoking a cigar, and snapping jokes out of the side of his mouth. Before he could play a note of jazz, white Chicago tenor saxophone star Bud Freeman involved himself in popular music by imitating Joe Frisco:

> I was a dancer; wherever I went, I did the Frisco, you know, the thing Joe Frisco got from Snake Hips Robinson [*sic,* Tucker], a dance called "the Frisco."

Variety noted that Frisco's imitators in vaudeville and the night clubs "have been as plentiful as were those of Eva Tanguay in her day and George M. Cohan in his. Some people have seen so many 'friscos' they believe it isn't a jazz dance unless the dancer has a cigar."[13]

Several female stage dancers also popularized a refined, stylish form of jazz dancing in Chicago. The Arsonia Cafe, opened in 1904 at 1654 Madison Street by Mike Fritzel, who had come to Chicago from Nebraska on a cattle train in 1898, and its reincarnation called Friars Inn, allegedly presented Joan Crawford (born and stage name "Lucille Le-

Seur.") Crawford later danced the Charleston in a 1928 movie about flappers called *Our Dancing Daughters*. Ginger Rogers also danced to the music of various pre-jazz orchestras. Bee Palmer, less well known today than the other two dancers, specialized in the Shimmie, the torso shaking dance step (borrowed from black cabaret dancers) which had separated "nice girls" from brazen flappers. *Variety* reported that her "refined shimmie act" was less vulgar than many other such turns:

> The undulating oscillations and nerve control of the involuntary muscles, particularly the pectoralis, major and minor are ... remarkable.

Gilda Gray, a Polish immigrant from Milwaukee who was streamlined with the help of vaudeville and cabaret vocalist Sophie Tucker, took a refined shimmie to Chicago and New York.[14]

The advent of prohibition in 1919 brought about further, sweeping changes in the early concert-saloons and cafes. Before the passage of the Volstead Act, "night resorts," as *Variety* called them, "were very little above the common saloon." Entertainment was merely something patrons got "for nothing along with the pretzels and the free lunch ... an old dilapidated piano, a couple of coon shouters, and girls ... to keep the customers in good humor and help them part with their sheckels." Prohibition required that saloons find something effervescent with which to camouflage the sale of alcohol. The solution most often taken was to upgrade the entertainment into a small vaudevill show. Since musicians were needed anyway, one inexpensive possibility involved featuring the band as a cabaret act. "Jazz, flappers, 'sugar daddies,' alias 'butter and egg men,' soon came to the fore and with them a complete rehabilitation" of saloons into cabarets."[15]

The earliest ragtime bands and demonstration dancers commercialized the popular dance craze in Chicago. Neighborhood clubs and fraternal societies with colorful names like the "Put Away Trouble Club," "The Merry Widows," "The Fleet Foot Dance Club," "The Dill Pickle Club," and the "Gladiators" hired bands and gave dances. In an age which had not yet supplied urban youth with playgrounds or organized sports, such small, obscure, popularly initiated dances were far too numerous to control, and new musical and dance styles continued to grow spontaneously from modest, neighborhood dance halls throughout the era.

By comparison, the commercialization of new urban night-life activities, as well as the need to respond to urban reform pressures, required the creation of a product that could appeal to a socially heterogeneous white market. Chicago's commercialized white dance halls of the late teens and twenties encouraged a variety of different social dance orchestras which played an important role in mediating the tensions between spontaneous cultural expressiveness and the patterned, supervised worlds of commercialized leisure behavior.

The city's pioneer in large, clean, well-lit, safe commercialized dance halls, presenting old fashioned but politely peppy dance music, was J. Louis Guyon, an early exponent of "clean" dancing whose moralistic approach paralleled that of the urban reformers. Guyon began organizing dances in Victoria Dance Hall beginning in 1909. In 1914, he leased the Dreamland Dance Hall at Van Buren and Paulina on the West Side, before opening in the same year Guyon's Paradise, which featured a dance floor said to accommodate 4,000 persons. Guyon reached for a mass market among the estimated 600,000 potential customers living in the West Side rooming house district; he located his Paradise dance hall near abundant, cheap late night transportation, and could draw customers from as much as five miles away.

Louis Guyon reportedly made a fortune running "the most conservative ballroom in Chicago." He stubbornly resisted the more daring and energetic developments in social dance and musical styles: before World War I, when the one-step and fox trot become popular, he allowed only the waltz and the two-step. He similarly banned the Charleston in late 1925, when it was imported from New York City. He catered to a somewhat older clientele than most of the ballrooms in Chicago. *Variety* characterized the music at the Paradise as "a service brand of dance music—peppy but not hotsy-totsy." The customers at Guyon's Paradise took their dancing seriously; social dancing was not just "one item of the evening, as in a cafe." Dance bands paid close attention to finding moderate but steady tempi and plenty of popular melodies.[16]

The biggest of the twenties "ballrooms," most of them built in the nineteen teens, offered the public carefully engineered and sanitized musical and dance experiences. While it was waging its fourteen-year war against "vice," the Juvenille Protective Association came to a culturally influential understanding with the owners and managers of nearly all of Chicago's largest dance halls, where some important jazz musicians earned a living. In 1921, the owners of the Midway Gardens, Merry

Gardens, White City Ballrooms, Trianon Ballroom, Marigold Gardens, Dreamland Ballroom, and the Columbia Ballroom formed the National Association of Ball Room Proprietors and Managers. Together, they pledged to work out their differences with the Juvenile Protective Association.

One step toward dance hall reform, taken in 1921, had a direct causal relationship to the birth of the Jazz Age. In an initial meeting with the JPA, dance hall managers had asked: "What can we do to make our dance halls more respectable?" The JPA's answer: "'Speed up your music.'"

> Within twenty-four hours, every orchestra in the ballroom group had doubled the tempo of its melodies. The toddle, the shimmy and kindred slow syncopated motions were impossible at the brisk pace the music set, and the managers found most of the bad dancing eliminated.

Second, ballroom managers had agreed to hire JPA observers ("hostesses") to supervise social behavior of the dancers. This chaperone system had "suggested itself as the outgrowth of the hostess devices employed during the war in army camps."[17]

Cooperation between dance hall entrepreneurs and urban reformers shaped the commercialization of the dance craze and created a demand for fast paced "peppy," but morally sanitary, jazz age social dance music. The Benson Orchestra of Chicago, one of the leading dance orchestras in a number of dance halls during the 1920s, specialized in such music. When this band recorded numbers like Jelly Roll Morton's "Wolverine Blues" with the influential Frank Trumbauer on C-melody saxophone, the musical gap between Chicago social dance music and jazz greatly narrowed. Even the Benson Orchestra's renditions of numbers like "Go Emmaline" and "San" set high musical standards for the performance of arranged-but-hot dance music.

James Lincoln Collier's description of New York City dance orchestras in the twenties aptly describes this influential Chicago counterpart. Dance bands like Edgar Benson's concentrated on a "smooth exposition of melody, with good tone, clean attack, [and] accurate execution, at the expense of improvised risk-taking." Groups like this one also sought to appeal to a broader range of sensibilities, recording waltzes ("Pal of My Cradle Days," "I'm Drifting Back to Dreamland," "Tears of Happiness") and a number of sentimental songs ("Fair One," "Lonely Little Wall-flow'r," and "After All, I Adore You") which were calculated to appeal to a brand of lingering Victorian sentiment against which Chicago jazz bands were in rebellion.[18]

JPA officers Jessie Binford and Elizabeth L. Crandall regularly inspected the leading dance halls. This arrangement worked well and kept the leading dance halls in operation for six years before the depression set in. In late 1928, however, Chicago newspapers discovered that the National Association of Ball Room Proprietors and Managers was paying at least one JPA official—Elizabeth Crandall—for her onsite investigations! Crandall's surviving reports on NABRPM dance halls were all favorable. Thereafter, JPA made an effort to keep expense accounts, which would help to justify the dispersal of funds collected by supervisors from those being supervised. The stock market crash and depression threw most of the dance halls temporarily out of business and left Elizabeth Crandall writing angry letters about expense payments due to her.[19]

Several of Chicago's large, commercialized dance halls which opened in the nineteen teens—dance halls subsequently associated closely with the Jazz Age—had gradually fashioned distinctive musical and dance formulae, choosing from among several degrees of dance music "respectability." In 1915, Patrick T. Harmon took over direction of Dreamland Ballroom on the near West Side, just outside the Loop at Paulina and Van Buren, where three elevated lines converged overhead to create what must have been a severe acoustical challenge. Harmon established a 1920s social dance and music polity "of moderate liberalism with the stop signal up on the too hot."

Harmon's Dreamland catered to a heavily working class neighborhood: "It is surrounded by families of the factory-working type. When the mechanics and [telephone] operators get washed-up after supper they have the urge to move to the strains of music." Harmon fully cooperated with urban reformers and focused attention on music and social dancing, despite the working class neighborhood in which his hall was located. Elizabeth L. Crandall of the Juvenile Protective Association praised the "well lighted and supervised" facility and emphasized the musical focus of the proceedings she observed there:

> The community singing here is always a unique feature and an interesting one. During intermissions a crowd gathers around the pi-

ano where Mr. Mehle leads them in community singing until the orchestra starts playing again for dancing.

Crandall even approved of the dancing, for although she witnessed "some eccentric dancing here," she saw even more of the "regular foxtrot and onestep than eccentric. The young people were dancing with much vim and zest."[20]

At his Dreamland Ballroom, which doubled as a roller rink, "Paddy" Harmon allowed some jazz. In order to appeal to a younger crowd than that which patronized Guyon's Paradise, he hired the all-black Charles Elgar Orchestra, a refined reminder of the South Side's cabaret world. South Side musician Willie Randall recalled that Elgar played "good *bona fide* ballroom music" at Harmon's Dreamland. "I don't say you would identify it and tag it as jazz ..."[21] Taking advantage of a promising situation, Elgar started with a small combo and with Harmon's support, he added a few musicians. By 1921, he enjoyed such renown among the dancers that Isham Jones, seeking to enhance his own reputation as a jazz-oriented white orchestra leader, challenged Elgar to a highly publicized "duel" of their nine-piece bands in order to grab some reflected glory.

Elgar built a versatile professional orchestra which played a wide variety of styles, ranging from the decorous waltzes, which it performed from behind a cluster of potted palm trees at black banker Jesse Binga's year-end parties, to concert wind ensemble literature, and some of the latest popular styles. Elgar adjusted his music to the market. But at Harmon's Dreamland he sometimes featured the improvised solos of the wayward, playboy jazz cornetist Freddie Keppard.

Charles Cook's all-black Dreamland Orchestra, another versatile concert and dance orchestra, which again featured cornetist Keppard, as well as the easygoing, golf crazy jazz clarinetist Jimmie Noone, followed from 1922 to 1928. Cook's band was "a fixture ... warm and popular with the steppers." The dance hall even broke through to some hot jazz, when, in early October 1925, Harmon hired Joe Oliver's jazz band to play for the annual dance of the Checker Cab company, whose employees were just the sort of streetwise savants to appreciate African-American jazz.[22]

During the same years, Harmon also owned the Arcadia Ballroom, an unpretentious working-class neighborhood dance hall on the North Side at Broadway and Wilson Avenue where he again presented dance bands organized by Charles Elgar and led by Darnell Howard, among others. Here, at the height of the Roaring Twenties, an observer witnessed the democratic power of the ongoing grassroots dance movement. A group of youngsters from the neighborhood, turned-out as "sheiks" and "shebas," generated spontaneous dance contests: "popularity here is based on the ability to twinkle an eccentric toe." Off in the darker corners of the hall, "where the stag line converge[s] with the young frails," the dancers competed for popularity through the originality of their steps:

> When the individualistic dance style pleases one of the opposite sex a team is formed, and they venture out on the main floor. Everybody seems intent upon drawing the limelight in their [*sic*] direction. Fashions in stepping are exceedingly versatile and varied.

The Arcadia presented no professional dancers like Joe Frisco and Bee Palmer: the neighborhood youth were proud enough of their own creativity that "the boys and girls would probably regard a team of professional dancers with disdain. And at that, very few professionals could imitate some of the dancing pulled by the Arcadia amateurs."[23]

Another pillar of Chicago's commercialized and reformed jazz age was White City, an amusement park named after the World's Columbian Exposition of 1893 and run by Herbert A. and Ernest L. Beifield at 63rd Street and Cottage Grove Avenue. It was located in an immense, sprawling rooming house district, near the El Station, streetcar lines, and Illinois Central suburban trains. White City had opened in 1905 and was subsequently attacked in 1909 by the Chicago Law and Order League for renting out its facilities to morally suspect traveling carnivals. White City management learned by 1914 to create more wholesome thrills based on fast physical movement on "The Flash," its roller coaster, in its bowling alley, and at a roller rink. Two dance floors, which could accommodate up to 5,000 dancers, carefully separated young and old dancers. The main ballroom had a cushioned dance floor, where a middle-aged crowd danced the more traditional steps. At the Casino, according to a JPA witness in 1917, the dancing was more "modern," and a "jazz band of five pieces played all the fancy dance music." In the Casino, in contrast to the ballroom, "most of the girls had their arms around the men's necks and in many cases the faces together."

By at least 1916, the new syncopated social dance music at White City Casino consistently drew a crowd of the more daring, fast moving white teenagers: young men in the Casino wore spats and drank liquor; the girls somehow seemed "hardly dressed" and wore "gaiters that once were white." In the twenties, the Beifields featured Sig Meyers's Casino Druids which included at various times Danny Altiere, Arnold Loyocano, Wop Waller, Benny Goodman, and many of the jazz-influenced musicians who worked at Midway Gardens as well. *Variety* likened their music to "a blast from a furnace," "full of sock." "It is a sheik and sheikess element Mr. Meyer [*sic*] and his lads appeal to. They use strictly hotsy-totsy numbers [and] the clerks and stenos do their stuff." The all-black dance bands of Charles Elgar and Charles Cook played long runs at the Casino, while the House of David Jazz Band and Jimmy McPartland's Wolverines under the direction of the slick dance band booker Husk O'Hare accepted the occasional job, too.[24]

The Midway Gardens, at 60th Street and Cottage Grove Avenue, was, along with Harmon's Dreamland and the Beifields' White City Casino, among the most important Juvenile Protective Association-approved dance halls which catered to a young working-class audience. This hall offered hot, jazz-influenced dance music played by a mixture of serious white jazz musicians and dance band professionals. This dance hall's history is an interesting example of the shifting economic and political foundations of cultural hierarchies in American music. Designed by Frank Lloyd Wright and opened in 1914 as an outdoor, summer concert garden for the National Symphony Orchestra, the Midway Gardens had offered "high class modern cuisine" and other refreshments. The Winter Garden, the indoor concert facility of this sprawling complex, was also opened for "devotees of the prevailing craze for modern social dancing," when not used by the National Symphony under the direction of Max Bendix. Just before World War I, private financial backing crumbled, and the Gardens were sold to the Edelweiss Brewing Company, which turned the high brow music center into a beer garden that now shared the name of a more notorious club several blocks away. Closed in 1923, this second Edelweiss Gardens was reopened the following year as the Midway Dancing Gardens.[25]

The Midway Dancing Gardens built its own identity by presenting dance music that was advertised as hotter and more daring than that offered by the nearby Trianon Ballroom. As a show business trade paper noted, located so close to the larger, more modern Trianon, the Midway Dancing Gardens' band "has to be good since the ballroom offers only dancing and charges $1 to get in." The concert shell, built for the National Symphony, focused attention on the music made by the dance orchestra through the large, acoustically advanced sound board behind the bandstand:

> Backed by a tremendous sound board, the musicians conform to the usual "Chi" tempo of mixing in numerous fast trots ... while maintaining a rhythm conducive to stepping. The orchestra has something of a reputation which seems securely based upon merit. They drew heavily on a Thursday, an off night.

Another reporter scornfully noted that the Midway Gardens attracted the lower social class whites: "the boys with the round hair cuts and the 'hotsie totsies' gather ... it was 'Corn Festival' week." *Variety,* the show business trade paper, considered it Chicago's ultimate Roaring Twenties dance hall:

> There is not a mature person to be found on the Midway Gardens dance floor. The youths in their inevitable high-waisted dark suits and with long funny hair and glossy with vaseline are perfect prototypes of the ballroom sheiks. The girls, most of them pretty and all of them endurance dancers, are dressed flashingly. In the daytime, their occupations probably range from dipping chocolates to taking dictation.

Jazz pianist/composer Elmer Schoebel led the orchestra through 1924 into 1925. *Variety* reported that "'Hot Stuff' with a fast tempo is the only sort of music that receives any encouragement." Schoebel hired the rhythm section from the now defunct white New Orleans Rhythm Kings jazz band—string and brass bassist Steve Brown, banjoist Lou Black, drummer Frank Snyder, and saxophonist Jack Pettis—to play in his Midway Gardens Orchestra, combining them with such dance band professionals as trumpeters Arthur "Murphy" Steinberg, Frankie Quartell, and reedman Art Kassel.[26]

Schoebel's hot dance band made a series of recordings in 1923, under such names as the Original Memphis Melody Boys, the Midway Gardens Orchestra, Midway Dance Orchestra, and the Chicago Blues Dance Orchestra. "Blue Grass Blues" and "House of David Blues" by the last-named of these bands were labeled by *Variety* as "typical examples of the colored indigo school of dance compositions," and these records give a good idea of what many white working-class dancers must have understood the term jazz to mean in 1923. The eleven-piece orchestra played ensemble-dominated, frequently polyphonic versions of original tunes written by Schoebel and other members of the band. The two trumpets usually took the lead in harmony, a common pattern among white dance bands of the twenties. The orchestra interjected plenty of two-bar instrumental breaks and favored cup-muted trumpet effects. The trombonist played in a solid New Orleans tailgate style, and the clarinet contributed excellent New Orleans obbligato embroidery. The reedmen doubled on saxophones, sometimes taking the melodic lead for a chorus, and otherwise playing harmonic and rhythmic support. Schoebel's head arrangements, riff figures, and movement of the lead from one to another of the wind instruments contributed variety and kept the polyphonies from becoming cluttered.

But the band's rhythm was not without anomalies. The rhythm section of Black, Snyder, Brown, and Schoebel swung steadily and happily along, laying down a smoothly infectious beat at medium dance tempos; banjoist Lou Black in particular played with a cocky assurance which propelled the rhythm into a jumping trot. But the modern sounding jazz beat was nearly neutralized by what was often a stiff syncopated ragtime phrasing by the trumpets. None of the band's lead horns, not even Murph Steinberg, who hung around with jazz musicians, had advanced rhythmically as far as the musicians who had come over from the New Orleans Rhythm Kings. To compound what amounts to a rhythmic confusion, the clarinetist played his breaks and solos in a legato, quarter- and half-note style which ignored the band's pulse.

The Midway Gardens Orchestra corrected its jazz shortcomings when, in 1926–27, dance band leaders Sig Meyers and Floyd Towne brought in the hottest young whites on wind instruments—intellectual looking reedman Frank Teschemacher, school dropout and jazz cornetist "Kid Muggsy" Spanier, and saxophonist Danny Altiere, as well as the slickly handsome Mississippi River valley pianist Jess Stacy. These musicians could share and even enliven the Midway Gardens dance beat with their urgent, rowdy, unpredictable instrumental work.[27]

In contrast to the Midway Gardens, Dreamland, and White City, all built before the war, the Trianon Ballroom at 62nd and Cottage Grove, built in the twenties and a NABRPM member from the start, sailed through the jazz age attracting large crowds of dancers with a conservative music policy which explicitly avoided hot jazz. Built in 1922 by entertainment entrepreneur Andrew Karzas for the then unheard of sum of $1,000,000, the Trianon was at that time the most expensive and elaborate of the commercialized dance halls built in Chicago to capitalize on the popular dance craze. Surrounded by Louis XVI-inspired decor, the dance floor was built to accommodate 3000 dancers, with room for an equal number in the many foyers, promenades, and loges. The Trianon was strategically located in the growing far South Side Woodlawn rooming house district for "working girls and laboring men" to whom it offered evenings of glamorous, urban sophistication. In addition to its dance music, the ballroom presented floorshows and demonstrations of aids to glamor, producing commercial dramas such as this: a woman, asked why she had come to the Trianon, replied that:

> She was hunting her husband 'Cy, who was in Chicago … she was going to be beautiful so that 'Cy would not neglect her and this was a cue which led to the opening of the curtains and the disclosure of a Marinello beauty parlor.

To young people recently arrived from the country, or to the children of immigrant parents, the Trianon might have seemed pretty "jazzy," but the dance music there was several steps removed from what insiders considered Chicago jazz. Karzas opened his ballroom with the Paul Whiteman Symphonic Jazz Orchestra at the royal sum of $25,000 for six nights of music, "the highest salary ever in the dance orchestra world." But the band flopped with Chicago dancers, who had come to the opening instead of going to their usual neighborhood dance halls, expecting a steady rhythm for dancing. But, as one of Whiteman's musicians confessed to local jazz entrepreneur Bert Kelly at the grand opening: "My God, they just wouldn't dance to our music."[28]

One probable explanation lay in Whiteman's focus on jazz for listening rather than for dancing. He championed "symphonic syncopation," and sometimes denied that his orchestra even played jazz: "What we have played is 'syncopated rhythm,' quite another thing. And our orchestrations have always been worked out with all the color and beauty of symphonies." This concept of jazz-influenced music-making later led to Whiteman's pioneering concert at Aeolian Hall in New York City on Frebruary 12, 1924, and encouraged complex musical arrangements made primarily for listening rather than dancing. Whiteman cleverly manipulated tonal colors and rhythmic patterns to maximize aural interest. He succeeded so well that many dancers could no longer find the steady, swinging Chicago beat which made dance hall patrons want to dance. As the New York *Clipper* put it: "Those who like his music refuse to patronize a dance hall and mingle with the masses; while dance hall patrons won't pay $2 to get into a Whiteman concert." Abel Green, one of the very early show business writers to follow jazz and dance music closely, put in:

> ... no dance organization can flit from the dance floor to the stage
> and do justice to either. A stage dance orchestra cannot play the
> fancy arrangements and really maintain a perfect dance rhythm.
> That's a musical impossibility.

Less technically oriented jazz bands often knew better how to please the dancers:

> In a dance hall, amidst hundreds of feet, the special arrangements
> are literally lost in the shuffle while the "hot" band with no pre-
> tense at symphonic qualities, blares forth the rhythmic jazz in a
> manner to please the masses.

In the twenties, large numbers of young Chicagoans, who followed the latest developments in social dance music, wanted "that steady hammering rhythm," and Chicago jazz musicians valued those who kept "good time."[29]

Although the Trianon was an important "jazz age" institution in Chicago—Karzas even brought in Rudolph Valentino for a one-night-stand—people who were looking for hot dance music would have scorned the waltzes which Karzas reportedly mandated every third number. They also would have resented the tuxedoed "floor men" who circulated constantly to reprove any who indulged in wild dancing and/or displays of affection on the dance floor.

The Trianon subsequently hired a series of dance bands led by such white leaders as Roy Bargy, Isham Jones, Paul Biese, Art Kassel, Dell Lampe, Arnold Johnson, and any of a number of dance orchestras booked by the ubiquitous Edgar Benson. These bands presented a more refined, old-fashioned dance music, catering to the sensibilities of an older audience for whom dancing to a gently bouncing beat in a public place seemed daring enough in itself. Lampe's Orchestra from the Trianon Ballroom, for example, nodded in the direction of hot twenties jazz when recording a warm version of "Prince of Wails" but demonstrated its concern for an older, more conservative crowd by waxing "Trianon Chicago Tango," "Lady of the Nile," and "The Midnight Waltz."[30]

The impulse behind the Trianon and its relationship to jazz were caught by band leader Meyer Davis, who sponsored a contest to rename "jazz" for people who disliked the word's lowlife implications. "Synco-Pep" won from a group of sanitizing euphemisms which included "Rhythmic-reverie," "Rhapsodoon," "Peppo," "Exilera," "Hades Harmonies," "Paradisa," "Glideola," and "Mah Song." Guy Lombardo and His Royal Canadians played extensively for Chicago dancers in the 1920s and sought to please the same upper-middle-class white audience as Meyer Davis. Discographer Brian Rust was surprised to discover that "when required" Lombardo's dance band of the twenties could "play as 'hot' as any of their contemporaries." "St. Louis Blues," "The Cannon Ball," and "Mama's Gone, Goodbye" startle the jazz fan with their stomping heat. But Lombardo was more likely to record waltzes like "Charmaine," "Ramona," and "Sweet Chewaukla, the Land of Sleepy Water" and a broad range of mawkish vocal numbers like "Sweet Dreams," "Please Let Me Dream in Your Arms," and "You're the Sweetest Girl This Side of Heaven."[31]

Other large, commercial, all-white Chicago ballrooms similarly refined, diluted, and camouflaged jazz. For example, Fred and Al Mann's "Million Dollar" Rainbo Gardens, at the intersection of North Clark Street and Lawrence Avenue on the Near North Side, featured a stage which opened onto the outside gardens and also moved inward to the center of the dance floor. The Rainbo Gardens staged "Elaborate spectacles and vaudeville numbers—ballets in the grand manner of the great opera houses" for which Frank Westphal (a novelty pianist in the Zez Confrey tradition who was once married to Sophie Tucker) and his Rainbo Orchestra provided the accompaniment in the early twenties. This band sounded much closer to jazz than most white social dance orchestras. They recorded "State Street Blues," "Oh! Sister, Ain't That Hot?," "Beale Street Mama," and "Aunt Hager's Blues."

But, according to *Variety,* the Rainbo Gardens appealed to "the white collar, middle class element ... family groups that would not feel comfortable in the hotsy-totsy environment of the cubby-hole cafes and can step out at the Rainbo and still be dignified." For these customers, Westphal recorded "Where the Volga Flows" and "That Lullaby Strain." Recordings made by the Victor Company of Ralph Williams and the Rainbo Gardens Orchestra show a more predictable adaptation of jazz numbers like Elmer Schoebel's "Prince of Wails" into a peppy medium tempoed arrangement lacking instrumental solos or insistent rhythmic heat.

Like Rainbo Gardens, the Eitel brothers' Marigold Gardens at the intersection of Broadway, Grace, and Halsted streets, was another sprawling complex of terraces, indoor and outdoor dance floors, and dining facilities. It featured elaborate stage revues, sometimes involving twenty-five to forty chorus girls, rather than hot dance music.[32]

Whenever ballroom proprietors moved toward a dine-and-dance or supper club policy, the music took on a more soothing, conservative cast. The Terrace Gardens, adjoining the Morrison Hotel at Madison and Clark streets, for example, was the largest dance hall in the Loop, designed to faintly echo jazz age sensibilities. The dance floor was surrounded with banked rows of tables and chairs so that as many as 900 customers could seat themselves between dance numbers and during floor shows and revues. The proprietors billed the Gardens as "Chicago's Wonder Restaurant" and presented featured attractions like "Elizabeth Friedman and her Famous Elida Ballet in a Musical Revue of Dance Divertissements." For social dancing, patrons of the Terrace Gardens stepped to a series of large, refined dance bands like those of Jimmy Travers, trumpeter Fred Hamm, saxophonist Paul Biese, and others under the direction of the Edgar Benson organization. They all provided a gently bouncing beat, violins, and plenty of predictable, reassuring melody. *Variety* concluded that Terrace Gardens catered to an "out-of-town clientele ... the man from Keokuk, Iowa ... transient middle class Rotarians."[33]

The show business trade papers and most patrons fully accepted the refined, bouncy dance music that was played in Chicago's large, racially exclusive hotel ballrooms as jazz. The hotel dance orchestra often defined itself in contrast to the "rackety, 'blarey' conglomerations of noise makers" of past years, searching for something "softer," with a "fullness of tonal quality ... decidedly more melodious and symphonic" with which to entertain the customers, who had paid at least a one-dollar cover charge which acted "as a sort of refiner keeping the undesirable element away."[34]

Playing in the leading hotels required a regimentation of behavior and appearance. Dance band musicians were expected to "be at all times immaculate in appearance, gentlemanly in conduct, clean-cut, well-bred, and business-like." One had to "realize his place and never attempt to assert authority." The orchestra took care to coordinate its activities with the *maître d'* to avoid interfering with the food service. One prominent leader recommended that hotel dance orchestras play 15-minute sets followed by 12-minute intermissions, allowing the customers time for five 3-minute numbers and then time to eat and relax.[35]

Out-of-town businessmen and tourists with plenty of money to spend could easily find jazz-influenced dance music down in the Loop at the Sherman Hotel's College Inn, where Isham Jones and his Orchestra were the sole attraction for many years. Jones, along with San Francisco dance band leader Art Hickman, had been among the first to abandon the early twentieth-century hotel formula of "Viennese violin ensembles," which had played behind rows of potted palms, for a featured dance band, which included choirs of saxophones. Jones provided a zesty brand of popular dance music in which the complex, harmonically advanced arrangements by Jones himself and his pianist Roy Bargy managed not to overwhelm the spirit of good times.

Jones' band was careful to "watch its step" in catering to "unsophisticated tourists and dignified Chicagoans." The College Inn was described as "gathering patrons of the better class of transients and among the wise crowd, it is oft slangily termed the 'sucker joint of the Windy City.'" According to *Variety,* the rowdy "soc" [sic] stuff was left to "working girls' ballrooms" like the Midway Gardens. Jones entertained a well-heeled clientele, who descended the stairs onto a long plush red carpeted aisle called "Peacock Alley," along which the royalty of Chicago's carriage trade, tourists, and celebrities strutted their stuff. In addition to the dollar cover charge the patrons paid "relatively expensive" prices for dinner.[36]

Isham Jones went much farther than any of the other Chicago hotel dance bands in casting African-American blues numbers within involved arrangements. His records of "Aunt Hagar's Children's Blues," "Henpecked Blues," and "Forgetful Blues" all show a sharp staccato ensemble attack and heavy vibrato from all the wind instruments. While the rhythm sections seem to mount some momentum, the solo horns seem not to break their solos into regular rhythmic units at all; cornetist Louis Panico, a self-professed admirer of King Oliver, turns mute techniques into musical humor, fashioning all sorts of unusual solo muted effects which, nevertheless, ignore the underlying rhythmic pulse. He and the clarinets reach regularly for satirically comical effects reminiscent of the Original Dixieland Jazz Band. When Jones took his dance band to New York, eastern audiences thought of his fast stepping numbers as "boisterous" "fireworks" from the "'sawdust' situation" of Chicago's Sherman

Hotel. California bandleader Vincent Lopez took over Jones' job at the College Inn in February 1925.[37]

From the end of World War I to about that same date, dance music in Chicago's other leading hotels was controlled by orchestra entrepreneur Edgar Benson, who fashioned exclusive booking arrangements with the Sherman, Drake, Edgewater Beach, Blackstone, LaSalle, and Bismarck hotels. Trade papers alleged that Benson created such exclusive bookings through kickback schemes; wealthy hotel patrons would hire his bands for private parties, and Benson "kicked back" to the hotel a percentage of what he earned.

Benson's power in the dance band business served to impede the passage of newer sounds into hotel clubs and to promote weaker dilutions of jazz. At one time or another, Benson booked the orchestras of Ben Bernie, Roy Bargy, Paul Biese, the Oriole Orchestra at Edgewater Beach, Arnold Johnson, Dell Lampe, Don Bestor, Frank Westphal, Jean Goldkette (before he moved to Detroit), and his own Benson Orchestra. These bands moved about, playing long-term contracted runs at the Green Mill Gardens, Trianon Ballroom, Marigold Gardens, Merry Gardens, the Pershing Palace, Crillon, Deauville, the Tent, Silver Slipper, La Boheme, Pantheon, the Senate, Tip Top Room, Rector's, the North American, the Midway Gardens, and Guyon's Paradise. As *Variety* put it: "It is next to impossible for any outside orchestra to come into Chicago at the recognized Benson trencholds ..."[38]

Benson's grip on "high class" white dance music perpetuated what trade papers considered a conservative brand of "symphonic syncopating combinations without the noisy effects produced by jazz bands." The elaborate arrangements appealed to a variety of delicate Victorian sentiments. When working for Edgar Benson, Biese, Bernie, and Bargy incorporated refined and restrained jazz numbers into their repertoires, relying upon a new breed of "specialty jazzers," well-trained dance music sight-readers, section men who could also take an occasional jazz solo. At the start of the decade, such musicians were rare; those sent by the union were too mechanical, "unable to follow the 'trick' stuff, or unacceptable of appearance."

When actual jazz musicians were willing to step in, the problem was solved, but many of them couldn't sight-read music or were scornful of the "sweet bands" and "likely to turn Bolshevik." A new breed of highly trained section men, however, began to master a tamed, dance band version of the jazz beat and solo inflections, often copying straight from jazz records. Louis Panico was the most celebrated of a group of trumpeters which also included Frankie Quartell.[39]

"Dance music technicians" developed their own ways of understanding the jazz elements expected of twenties dance bands: according to *Variety,* legitimately trained musicians in the jazzy dance bands thought of "jazz" as "snappy, but not necessarily in tune"; to these legitimately trained instrumentalists, "get hot" signified "scorching the air by blowing notes off key"; they defined a "hokum" player as "not necessarily a good musician but a good performer."

The surging jazz beat presented established dance band musicians with major problems. Most simply could not catch it. In order to assimilate what they understood about the new beat, dance band leaders hired college men, since "they have the spirit of the new generation in their dance rhythms." Many young college men were not only technically well-schooled enough to play the charts, but also in touch with the latest dancing rhythms. College musicians were more likely to have "the bearing and appearance of a gentleman," important when hotels kept a sharp eye on their employees (particularly musicians). They were young, handsome, and knew how to approximate enough rhythmic swing to appeal to the younger dancers. They could earn more—usually over $50 a week—with a major dance band than they would have commanded at most other jobs. It took them some time to realize that they were earning substantially less than a dance band's regular musicians.

Where Benson failed to gain control of a dance band or a booking, he could usually manipulate selected musicians in the orchestra. He booked individual players as well as groups on a commission basis and a "play-or-pay guarantee" under which the musician was assured of payment whether he worked or not, but could not control where or to whom he was assigned. Pianist/leader Roy Bargy was widely admired by jazz musicians and had been Bix Beiderbecke's tutor in harmony. Benson actively impeded Bargy's ambitions when he sought to break away on his own; Benson split up Bargy's band by withdrawing key musicians. He did the same thing to Isham Jones, luring away his star trumpeter Louis Panico and setting him up with his own band in vaudeville and at Guyon's Paradise. When Jones went into New York's Cafe de la Paix cabaret on Broadway without his star attractions, the club bounced his paychecks.[40]

Edgar Benson's power also depended on close cooperation with music publishers, retail music merchandisers, and recording companies. The social dance craze turned publishers of sheet music and orchestral arrangements away from vaudeville, where they had used vocal acts to promote their products, and toward prominent dance bands. Publishers tended to blame jazz for an ongoing slump in sheet music sales, concluding that jazz signaled a decline in musical literacy. As the New York *Clipper* put it: "A well written melodious number is so buried under the jazz antics of scores of orchestras that the audience can scarcely recognize the melody...." Benson's orchestras featured plenty of unadorned melody, thereby attracting the keen interest of the music publishing houses. Together, band leader and publisher arranged with recording executives to wax tunes created for the mass market; in fact, companies like Brunswick and Victor reserved potential hit songs for their popular recording orchestras, leaving the novelty tunes and ethnic music to little-known bands. Once a likely hit had been recorded, promotional activity could become intense. Well-known recording bands appeared in vaudeville, cabarets, hotels, and even music stores, where the windows were decorated with special publicity streamers, hangers, folders, catalogues, and cardboard cut-outs.[41]

In 1924 and 1925, however, several of Edgar Benson's most famous bandleaders began to follow Isham Jones in rebellion; first Paul Biese, then Roy Bargy, and finally even Don Bestor declared their independence. As a result, big dance bands—Abe Lyman's from Los Angeles, the Coon-Sanders Nighthawks from Kansas City, and New York's Vincent Lopez Orchestra—began to play long runs in Chicago's poshest dine and dance clubs. Moreover, a new breed of band bookers with national ambitions stepped in to promote the dance bands and jazz groups which Benson had shunned. Ernie Young, among the first of this influential group, had produced black-and-tan revues in some of Chicago's hottest cabarets and promoted Roy Bargy, Paul Biese, King Oliver's Jazz Band, Elgar and his Champion Colored Orchestra, the Seattle Harmony Kings with hot cornetist Wild Bill Davison, and the Marigold Garden orchestra. Dance band leader Ray Miller booked Ben Bernie and his Notable Dance Orchestra into New York's Roosevelt Hotel; he and I. Jay Faggen organized Cosmopolitan Orchestras Booking and Promotion which handled King Oliver's Jazz Band and vocalist Red McKenzie's Mound City Blue Blowers.[42]

Especially important for the spread of a more heated brand of dance music by both white and black jazz bands was the Music Corporation of America, organized by Jule C. Stein and Ernie Young in 1922. At first locked out of Chicago by Edgar Benson, MCA subsequently moved in to precipitate his decline, booking many of the hotter bands which appealed to the younger dancers. Stein was also the first to effectively organize national booking of dance bands and dance halls, opening a New York office in 1926 in order to better book MCA bands on the east coast and particularly in New England. Stein reportedly built an elaborately equipped, experimental ballroom near Waukegan, Illinois, where his production department worked to develop the "entertainment possibilities" of the many local bands which vied for broader exposure.[43]

In terms of sheer numbers, therefore, more Chicagoans experienced the jazz age in large, racially segregated, commercialized dance halls and hotel supper clubs than in the jazz cabarets. The music which these institutions presented varied according to the sensibilities of the proprietors and managers, but represented an attempt to assimilate some jazz within a broader range of tastes. The Midway Gardens, White City Casino, Harmon's Dreamland Ballroom, and Harmon's Arcadia played major roles in popularizing several styles of jazz-influenced dance music.

In these dance halls, a jazzman like Freddie "King" Keppard, relinquishing his role as a dance band section leader, might take a sip from the straw which he had inserted into a bottle hidden in his inside jacket pocket, and shortly thereafter break into a searing solo flight which would momentarily transform a more restrained evening of dance music. He, Louis Armstrong, Jimmie Noone, and young, innovative white jazz musicians like Frank Teschemacher and Benny Goodman stabbed out the occasional driving, unpredicatable solos which added the heat of Chicago's tough, fast-moving, youthful sensibilities to Jazz Age dance music.[44]

But several of Chicago's white jazzmen, musicians who devoted their lives to jazz, often felt constrained by the straight melodic or harmonic ensemble lines in what they called the "sweet" dance bands. Their fiery solo sensibility could be expressed more freely in a smaller orchestra and a less regulated venue. Famous jazzmen like Keppard, Spanier, Noone, Stacy, Goodman, and Teschemacher maintained long working relationships with the dance bands, but other devotees of hot rhythms, polyphonic ensembles, and improvised instrumental solos denigrated the dance orchestras as a commercial compromise of their artistic freedom. Their careers in jazz were often grounded on their efforts to turn their understanding of South Side Chicago jazz against the commercial compromises of the "sweet" dance bands. It is time to look more closely at the relationships of the white Chicago jazzmen to the cultural forces that molded Chicago in the twenties.

4

White Chicago Jazz: Cultural Context

JAZZ QUICKLY BECAME a major musical expression of white sensibilities. Among the factors contributing to this were the performances of white New Orleans bands beginning in 1915, the early appearances of black orchestras in white clubs and dance halls, the assimilations of ragtime and nut jazz by white Chicago dance bands, the jazz band records of several white groups recorded in New York City, the popularity of African-American musicians in the South Side black-and-tan cabarets, the pioneering jazz and jazz-influenced dance hall music of Elmer Schoebel, and, lastly, the determination of some white Chicagoans to take inspiration from South Side musicians and become jazzmen, too. By the mid-twenties, an inner circle of brash young whites began to appear, who strongly identified with Chicago's jazz scene and devoted themselves with a religious fervor to jazz music and the jazz life. Their careers began five to ten years after South Side clarinetist Darnell Howard appeared with Wickliffe's Ginger Orchestra, "America's Greatest Jaz Combination," in 1916.[1]

In twenties Chicago, white jazz musicians labored under a considerable disadvantage among knowledgeable cabaret goers. They were relatively inexperienced when compared with the South Siders. According to

Paul Eduard Miller, an early chronicler of Chicago jazz, the relatively small inner circle of acute jazz listeners in the 1920s recognized that black musicians played better, more mature, more confident jazz than whites.[2]

The general public, although less aware of the differences between good and less good bands, also tended to think, because of the long traditions of minstrelsy and vaudeville, that blacks played "real" jazz. As a result, the aspirations of white Chicago jazzmen were caught between the reputation of the black stars in the South Side cabarets and the demonstrated ability of the white dance bands, which were so well ensconced in the large hotel ballrooms and the dance halls. Would-be white jazzmen couldn't become black, although one of them—Mezz Mezzrow—tried mightily. They could, however, learn to play hotter and wilder jazz than that offered by the white dance bands. Some rebellious young Chicagoans cultivated a strong sense of cultural alienation from the suburban middle class, turning their experience of South Side jazz and their understanding of racial injustice into a critique of mainstream America.

The impulse to play a wilder, freer music arose not from race but from the excitement of urban life. Like the black musicians, those whites who determined to play jazz had been deeply stirred by the excitement of Chicago, itself. Early twentieth-century Chicago produced a generation of teenagers of all races whose lives were more deeply influenced by the experience of the city than by traditional rural definitions of family, culture or career. Many such youngsters worked as newspaper sellers, bell hops, telegram runners, saloon entertainers, messengers, advertisers, incidental laborers, and street vendors of all sorts of merchandise, from chewing gum to cigarettes. Kids like these sometimes worked or played late into the night, ate whatever and whenever they pleased, rarely attended school, might smoke and drink from very early ages, and sometimes used drugs like marijuana, which was then sold openly on Chicago street corners. The music of the white Chicago jazzmen—the McPartlands, Bud Freeman, Mezz Mezzrow, George Wettling, Muggsy Spanier—expressed some of the tough freedom of such youngsters, who lacked formal education and were powerfully drawn to what Jane Addams called "The Spirit of Youth and the City Streets."[3]

Urban reform observers of the mid-twenties street cultures asserted that such youth were drawn to the thrill of urban freedom; a typical youngster learned "to like and to depend upon the excitement of the streets and he sees all aspects of vice ... he acquires a distaste for a regular occupation." One Chicagoan recalled newspaper boys from ten to fifteen years old, for example, as they roared off into the night in the back of delivery trucks:

> The joy of it. They cling on casually, waving unconcerned arms and legs at the sides. They disappear into the night. Riding on swift going trucks at night. Free of control. Yet controlled for the sake of [newspaper] circulation ... what home could compete in fascination with the lure of those night rides on the rushing trucks, that undisciplined, nomadic life which is recruiting ground for our gangs and our criminals?

Similarly, "Children of the Stage," very young ones, became veteran actors and performers, like the two "mere infants" who did "an Amazing Apache dance" in dozens of Chicago theaters, cabarets and night clubs, performing until 2:00 a.m. Such youngsters feasted on urban popular entertainment, eagerly mixing in the itinerant street carnivals and backroom gambling parlors, spending long hours in the vaudeville theaters and cheap nickleodeons, hanging around at the swinging doors in hopes of earning some change by entertaining the habitués with a song. These resilient ragamuffins read "the small magazines like Red Pepper, Zip, Jazz, Whiz Bang, Frolics, Short Stories, Hot Dog, and the Wampus Cat." They knew where to find excitement, discovering everything from obscene post cards for sale in seedy store fronts, to gangster hideouts, and wild apartment rent parties with their tough, syncopated blues, ragtime, and jazz.[4]

Upper-class urban reformers probably exaggerated the lower-class basis of these urban street sensibilities, which also, according to the novels of James T. Farrell, seized the high-school age children of the white middle class on the Far South Side. Bored by school, repulsed by and somewhat afraid of organized religion, angry at authoritarian parents who remained determinedly out of touch with the youth culture of the streets, Studs Lonigan and his friends search for adventure in the defiance of middle-class respectability by running in gangs, fighting, stealing, smoking and chewing tobacco, and discovering the forbidden pleasures of city life.[5]

The white Chicago jazzmen who thrilled to the excitement of twenties Chicago hailed from both working- and middle-class families and came from a variety of ethnic backgrounds. Milton "Mezz" Mezzrow,

who was born on the Northwest Side in 1899 and who wrote the best
known of the white Chicago jazz autobiographies, clearly identified his
need to make music with a youthful sense of roaming the streets "all
jammed up full of energy, restless as a Mexican jumping bean ... I felt
like I wanted to jump out of my skin, hop off into space ..."

> It was a lot more than a mere sex flash that kept me all keyed up. I
> was maneuvering for a new language that would make me shout
> out loud and romp on to glory. What I needed was the vocabulary.
> I was feeling my way to music like a baby fights its way into talk.

Mezzrow ignored his parents' urging toward solid professional respectability. He hung around Division and Western avenues with a gang of
young toughs, who tried to "live our whole lives out before the sun went
down." Similarly, Chicago percussion instructor Roy Knapp, who taught
most of Chicago's leading white jazz drummers, remembered that Gene
Krupa, his star pupil, was a wild bundle of adolescent nerves, "drumming on the walls," when he first came to see about taking lessons.[6]

Jazz fanatic and chronicler Ralph Berton, brother of the noted drummer Vic Berton, proudly explained to his hero, cornetist Bix Beiderbecke,
that he had graduated from Goudy Elementary School, Winthrop Avenue at Foster, despite having spent a grand total of two years going to
classes. Vic and his other brother Gene grew up "outside the walls of
school and church ... complaisant parents ... spared [us] other sacred
institutions we saw imposed on all other kids...." Ralph Berton spent as
much time as he wished "learning, or doing, exactly what I wanted,"
which, in his case, meant exploring the city, reading at random, and
listening to records.[7]

Chicago's excitement also worked powerfully on trumpeter Max Kaminsky, who later remembered "the bursting feeling of life in the city,
engendered by the hordes of eager young people from Midwestern farms
and small towns who swarmed into Chicago by the thousands." Chicago,
"this bubbling, yeasty stew," was "a fast city [with] a fast life, a 'toddlin'
town.'" Kaminsky, born to Russian immigrant parents, in Brockton, Mas-
sachusetts, arrived on the Chicago scene in 1928, and found a world of
difference from the east coast. His colleague, banjoist Eddie Condon, did
a great deal to promote the notion of an inner circle of true believers,
whom he labeled in 1927 "the Chicagoans," and later "the barefoot mob"
or the Condon "Gang." They were jazz zealots who played a definable

style called "Chicago Jazz." Condon couldn't get into Chicago night life quickly or deeply enough, eagerly fleeing the repressive morality of rural Indiana and Illinois. "A bantam sized, cocksure little guy, with an impudent Irish face, a quicksilver mind, and a lethal tongue ... bow tied, debonair, tight-mouthed, and gimlet-eyed," Condon came fully alive only in the fast-moving world of jazz and cabarets.[8]

Few of these white Chicago musicians reasoned their ways to jazz; rather, they were drawn to visceral, nonverbal forms of personal expressiveness as teenagers or very young men. Their shared devotion to jazz functioned as a religious faith, providing a new form of sectarian cohesion and an idealistic scale of values. Jazz, as Neil Leonard has explained, often acted in this manner as a religion for many who had been cast adrift in the maze of urban America.[9]

This sense of defying the dull and predictable lives prepared for them by parents and school, of mutually discovering and shaping a powerful and esoteric medium of musical and personal expressiveness, particularly animated "the Chicagoans," a self-proclaimed inner circle of white true believers in jazz, surrounded, as they saw it, by a deadening middle-class world of crass commercialism. The label suggests an exclusive claim on the creative spirit of twenties Chicago jazz, but it has usually been applied to a relatively small group of white jazzmen who worked closely with "the Austin High Gang" and promoters Eddie Condon and William "Red" McKenzie. These jazzmen absorbed an enduring wild streak of urban musical primitivism and would remain stubbornly loyal to it; but even among Chicago's white jazz musicians of the twenties, they should be understood to have been just one especially influential wing of a larger group who merit the label "Chicagoans," too. This more numerous group of white Chicago jazzmen included musicians who shared the urge to make wild, exciting urban music, but whose sensibilities also allowed them to play jazz in technically more arranged and commercialized musical forms. The label camouflaged many white Chicago jazzmen—clarinetists Benny Goodman, Bud Jacobson, and Voltaire De Faut, drummer Vic Berton, pianist-bandleader Elmer Schoebel, for example—who played jazz band improvisations, read dance band arrangements, composed, arranged, and made substantial contributions to Chicago jazz in the 1920s.

As traditionally applied, the label "the Chicagoans" also has led to misunderstandings about who actually played most of the jazz in twenties Chicago. Compared to many of the South Side musicians, who never claimed to represent the city or its culture, many of the whites—significantly the "Austin High School Gang"—had short performing careers in Chicago during the twenties, although they did have longer careers later elsewhere. Leon "Bix" Beiderbecke was often associated with the city, but actually played relatively little there. Some important white jazz musicians, like clarinetist Bud Jacobson, alto saxophonist Boyce Brown, and reedman Rod Cless, slightly younger than the more famous of the white Chicago jazzmen, played long runs in Chicago clubs during the late twenties and throughout the 1930s, certainly earning their rank as "Chicagoans," even though their careers reflected less of the glamour of the 1920s of Al Capone, Big Bill, and Elliott Ness.[10]

All of Chicago's white jazzmen were inspired by their visions of freedom and excitement in the city, but their music, and especially their careers, moved in different directions, depending upon their approaches to resolving the contrary pressures arising from the influence of South Side music and from the legacy of the commercial white dance bands. Some white jazz musicians, best exemplified by Mezz Mezzrow, took an inflexible stand in favor of urban musical primitivism as opposed to the commercial "sweet" dance bands. Others, led first by Elmer Schoebel and later by Benny Goodman, found ways to express big city excitement in musically more sophisticated (and ultimately more lucrative) ways.

One can identify three fairly distinct subgroups within the most inclusive definition of white Chicago jazzmen, each with its own social origins: first, those white jazzmen—Benny Goodman, Muggsy Spanier, Gene Krupa, Art Hodes, Mezz Mezzrow, Floyd O'Brien, Volly de Faut, Joe Sullivan, Vic Berton, and Joe Marsala—who were born and/or raised in the city neighborhoods; second, those like Dave Tough, Frank Teschemacher, Jim Lannigan, the McPartland brothers, Bud Freeman, and Bud Jacobson, who were either born or raised in the suburbs; third, a larger group—Eddie Condon, Wild Bill Davison, Bix Beiderbecke, Hoagy Carmichael, Rod Cless, Frank Trumbauer, and Elmer Schoebel—who came to Chicago from various points in the Midwest in order to pursue their goal of becoming jazz musicians.

Several of the important white Chicago jazz musicians were raised in Chicago's relatively poor inner city neighborhoods. The largest number were from the Near West Side, with a sprinkling from the Far South and the Near North Sides. The Near West Side, an area bounded by Kinzie Street on the north, the Chicago River on the east, and the Burlington railroad tracks at 16th Street on the south, was the classic immigrant melting pot of first- and second-generation Italians, Germans, Polish and Russian Jews, and French Canadians. During the 1920s, Italians were the most numerous ethnic group on the Near West Side. That neighborhood was also famous for Jane Addams Hull House at Halstead and Polk streets, for numerous Italian gangsters ("Bootleggers' Square" was at 12th and Halstead), and for the dark and dirty alleys of inner city poverty. It produced clarinetist Benny Goodman, pianist Art Hodes, reedman and organizer Mezz Mezzrow, cornetist Jimmy McPartland, and probably tenor saxophone stylist Bud Freeman, too, although he and McPartland always preferred to talk about their subsequent years in Austin on the Far West Side. Drummers Gene Krupa and George Wettling and clarinetists Volly DeFaut and Don Murray grew up on Chicago's Far South Side, a white rooming house neighborhood, while trombonist Floyd O'Brien grew up in Hyde Park and cornetist Muggsy Spanier, who in 1924 led the Bucktown Five in the first significant jazz recordings in Chicago by white musicians, came from the Near North Side.[11]

Attitudes toward music varied according to socio-economic status, personal proclivities, and whether or not drinking had become an important ritual of life; still, certain important similarities in outlook characterized the poorer, inner city Chicagoans. Benny Goodman was one of twelve children born to Russian immigrants who, while moving around, settled for a time at 1125 Francisco Avenue in the Jewish ghetto on the Near West Side. Goodman showed less adolescent rebelliousness than Mezzrow and some of the others. While stirred by the restless sensibility of the city, Goodman saw music as a challenging body of knowledge, a craft, and a more creative avenue than factory labor for making much-needed money to support himself and his family. He earned his first five dollars when he was only twelve years old, standing up in the pit of the Balaban and Katz Central Park Theater to play an imitation of clarinetist Ted Lewis. Although his recorded improvisations conveyed plenty of fire, Goodman considered his music something "in my hand," a craft, an avenue of opportunity, a complex body of knowledge. He expressed a sense of professional distance from what he called "the wild West Side mob," and built a career in the recording studios and with the swing music of the 1930s.[12]

Pianist Art Hodes, whose parents had emigrated from Russia in 1904, concluded from his early experiences "on Chicago's tough West Side" that music represented, among many other things, economic opportunity.

After piano lessons at Hull House, where Benny Goodman also studied for a time, Hodes at nineteen went to work from 9:00 p.m. to 4:00 a.m. at the Rainbow Cafe on West Madison Street. "I was able to buy a car, wear nice clothes and walk around with money in my pocket all the time." Like Goodman, Hodes spoke of his music as a craft, and showed little of the wild, urban bandit characteristics cultivated by some white Chicago performers like Condon, Davison, and McPartland.[13]

Francis "Muggsy" Spanier shared Hodes' unassuming modesty about making music. Born as one of ten children and raised on the Near North Side at 117 E. Delaware Place, Spanier grew up loving boxing, baseball, and music. Nicknamed for John J. "Muggsy" McGraw, then manager of the New York Giants, he played in a street band when only six years old and supported himself as a messenger boy on LaSalle Street. Like all of the white Chicagoans, he paid little attention in class, immersing himself, instead, in various new forms of popular culture. Yet Spanier never posed as a social or cultural rebel. He worked for many years in the "sweet" dance and stage bands of Art Kassel and Ted Lewis with little apparent artistic frustration.[14]

Although their paths did not often cross, Gene Krupa shared Spanier's musical professionalism. Born into a large family of Polish-American Catholics who lived in a Far South Side neighborhood, Krupa channeled his nervous energy into the craft of dance band and jazz drumming. He smoothly negotiated potential conflicts with his parents about his preference for drumming over the priesthood. Krupa worked with the dance bands of Leo Shukin, Joe Kayser, and Thelma Terry, as well as the Seattle Harmony Kings, the Hossier Bell Hops, and the Benson Orchestra, becoming known as "cleancut, intelligent, shy, serious, and very ambitious." Much the same mixture of jazz excitement with musical professionalism characterized clarinetist Volly de Faut, who played both in important small groups like the New Orleans Rhythm Kings and the Bucktown Five and in larger dance bands such as those led by Sig Meyers and Ray Miller.[15]

So too, percussionist Vic Berton, born in Chicago in 1896 to a vaudeville theater violinist, took music as a craft and a profession, studying with Joseph Zettelmann, the first percussionist with the Chicago Symphony under Frederick Stock. Berton appeared with the Chicago Symphony and the Milwaukee Symphony orchestras when he was still only sixteen, and thereafter worked in the dance bands of Paul Biese, Arnold Johnson, Husk O'Hare, Art Kahn, and Don Bestor and also led his own band at the Merry Gardens. In 1924, Berton took over the management of Bix Beiderbecke and the Wolverines and occasionally played with the band, as well.[16]

Several of the most prominent inner-city Chicagoans, therefore, subsequently stressed the musical and economic (rather than the socially rebellious) dimensions of their lives in jazz. Among their published recollections, only reedman Mezz Mezzrow's best-selling, flamboyant autobiography *Really the Blues,* first published in 1946, strongly associated white Chicago jazz with social rebellion. He spoke of a "social revolution simmering in Chicago ... a collectively improvised nose-thumbing at all pillars of all communities, one big syncopated Bronx cheer for the righteous squares everywhere." But Mezzrow was an inner city Chicagoan born to a financially secure middle-class family. An uncle owned a chain of Chicago drug stores. Mezzrow's comparatively wealthy background distinguished him from the other inner city Chicagoans like Goodman, Spanier, and Krupa. He found a positive resonance among Chicago jazzmen for his ideas of rebellion through syncopation, improvisation, and blue notes.[17]

The second group of white Chicago jazzmen, the "Austin High Gang"—cornetist Jimmy McPartland and his guitar-playing brother Richard, tenor saxophonist Bud Freeman, clarinetist Frank Teschemacher, pianist Dave North, and string bassist Jimmy Lannigan—played an important role on the white Chicago jazz scene and a pre-eminent role in the subsequent critical definition of "Chicago Jazz." Although they performed relatively little in twenties Chicago and made their reputations as "Chicagoans" later in New York, they did form the core of the Blue Friars, who played on radio station WHT (Mayor *W*illiam *H*ale *T*hompson) in 1924 as Husk O'Hare's Red Dragons. In 1925, they were also members of Husk O'Hare's Wolverines, who played in Iowa and at the outdoor musical shell of White City Amusement Park. Several of them starred in some of the better white Chicago recording sessions—McKenzie and Condon's Chicagoans, Charlie Pierce and His Orchestra, and the Jungle Kings.[18]

Austin was an unlikely source of Roaring Twenties jazz. A comfortable, middle-class suburban-style community which had been annexed to Chicago in 1899, it was built by Henry Austin, a state legislator who drafted the Illinois Temperance Law of 1872. The town named after him was a center of the Prohibition movement, and saloons were conspicuously absent in the Twenties. Like a true child of the inner city, Mezz

Mezzrow described Austin as "a well-to-do suburb where all the days were Sabbaths, a sleepy-time neighborhood big as a yawn and just about as lively, loaded with shade trees, clipped lawns and a groggy-eyed population that never came out of its coma except to turn over."[19]

But two of the leaders of the Austin High Gang—Jimmy McPartland and Bud Freeman—were more closely tied to Chicago's neighborhoods than their many recorded interviews would indicate. Moreover, none of the leading Austin High jazzmen actually graduated from that school. Not one of them appears in the school's yearbook. By their own testimony, most of their school hours were spent at the town soda fountain, the Spoon and Straw at 5619 Lake Street, owned by C. S. Lewis, where they first heard the records of Ted Lewis, Paul Whiteman, Jan Garber, the Original Dixieland Jazz Band, and the New Orleans Rhythm Kings.

Bud Freeman, arguably the most consistently creative soloist of the group, admitted that he was obliged to go to a YMCA summer school in order to progress from Nash Grammar School to Austin High.[20] Statements in his recently published autobiography not withstanding, Freeman admitted in another interview that he had been born somewhere in downtown Chicago, not Austin. He was unable to recall more exactly where he had been born within the city, despite the fact that he had lived there for six years.

Jimmy McPartland was nearly as reticent about his earliest years, talking repeatedly about his Austin High School experiences. But Benny Goodman recalled having met McPartland not at Austin High but at Harrison High School in the poverty-stricken Near West Side. In a Smithsonian Institution interview, McPartland remembered having spent his first five years growing up in a poor, racially-mixed neighborhood on Lake and Paulina streets. When his parents temporarily divorced in 1912–13, McPartland was consigned to an orphanage and felt abandoned, "mad at everyone," and bitter; he later joined a street gang, and built a reputation emulating his father, a two-fisted scrapper. In 1919, when Jimmy was twelve, a court urged his reunited parents to move the family to a better neighborhood; they settled in Austin, where McPartland learned to make apparently "lawless" music as a substitute for a life of crime. "Jazz supplied the excitement we might otherwise have looked for among the illegal activities which flourished then in the neighborhood."[21]

More typical of the suburbs were middle-class people for whom rebellion through jazz served to channel and express frustrations stem-

ming from unhappy childhood experiences. Clarinetist Frank Teschemacher died before telling his story, but Bud Freeman has painted a
portrait of a shy, withdrawn, acerbic, troubled youngster completely
immersed in his jazz.

Oak Park drummer Dave Tough started hanging around with the
Austin High School musicians in 1922. Born to immigrant Scottish parents (his father was a bank teller who dabbled in the real estate and
commodities markets), Tough grew up reading voraciously. His mother
died in what is variously described as an alcoholic or apoplectic fit when
he was nine. Tough is said to have taken up serious drinking at fourteen.
After drifting away from classes at Oak Park High School, the young
drummer took language and literature courses at the downtown Lewis
Institute, and hung around at a cabaret called the Green Mask, where he
"accompanied" poetry readings by Max Bodenheim, Langston Hughes,
and Kenneth Rexroth. He learned a strong traditionalist's respect for
language and tried to base his life on jazz, literature, and alcohol. Despite
his supremely assured, rhythmically "relentless" drumming and flashing,
literate wit, Tough was painfully insecure, constantly searching for approval, his rebellion torn between the different, if related, sensibilities of
twenties' literature and jazz.[22]

The relatively comfortable, middle-class dimension of the Far West
Side of Chicago made a major contribution toward the musicians consid-
ering jazz as art. The Austin High Gang, simply by living a suburban-
style life, were more likely to come into contact with contemporary
notions of jazz and art than the very poor whites and poor blacks of the
inner city. In the suburbs jazz sometimes served to express artistic alienation from middle-class materialism. Dave Tough was the only one of the
white Chicago jazzmen who was active in the intellectual currents of
the time; the rest, however, were aware that their involvement in jazz
paralleled certain powerful intellectual trends. Tough worked hard to
convince the other white musicians that jazz expressed artistic sensibilities. He taught Bud Freeman and Frank Teschemacher about George
Jean Nathan and H. L. Mencken's *American Mercury,* Chicago's Art
Institute, and current thinking about artistic rebellion against the bour-
geoisie.

The greater number of the third group of white "Chicagoans," those
who migrated to the city from the Midwest and beyond, shared the
rebellious attitudes of the Far West Side "Chicagoans." Several of them,
far less intellectual than Dave Tough, enjoyed the mixture of artistic

sensibilities with jazz, even if they remained much better informed about jazz than art or the intellect. Max Kaminsky, who rarely read anything but the newspapers, noted that "all the fellows in the Chicago crowd," musicians with whom he retained important ties, were "on a genius kick." The cornetist admitted that until meeting them, he had "seldom cracked open a book," and Tough mocked his poor reading habits, but "... I didn't let it bother me too much for I always felt there were other ways to learn besides from books."[23]

The notion of social rebellion through jazz was a sensibility that the suburban jazzmen shared with several members of the third group of "Chicagoans," those who had migrated to the city from the greater Midwest. Just as Chicagoan Jimmy McPartland had felt abandoned and angry, so too had cornetist William Edward "Wild Bill" Davison found his life in Defiance, Ohio, hard to accept. Abandoned by his parents, he was brought up by grandparents who worked as librarian and custodian of the Defiance Public Library. Living in the library's basement, Davison, continually hushed by his guardians, grew up to force his rage through a cornet. He practiced in a row boat, drawn far enough out into the Maumee River that no one could tell him to quiet down. He plunged eagerly into the musical night life of Chicago. There, the raging fury of his cornet had no equal.[24]

Davison discovered an immediate empathy with fast-talking, sardonic Eddie Condon, whose promotional abilities earned him the nickname "Slick." Condon's family moved to Chicago via Goodland, Indiana, Momence, Illinois, and Chicago Heights, as his father, a saloon-keeper, steadily lost ground to the harassment of the Women's Christian Temperance Union. A rhythm ukulele, banjo, and lute player, Condon inherited a strong sense of Irish-American alienation from midwestern Anglo-Saxon Protestants. He retained the attenuated ties of most of the white Chicagoans to the intellectual and artistic dialogues of the day. A third-generation Irish-American, and a tough survivor who cultivated a nonchalant hedonism, he admitted that after suffering through some very, very literary conservations with college student fans, his own reading had never included Marcel Proust.[25]

Whatever the differences between the three groups of white Chicago jazzmen, they all passed through a complex process of personal orientation toward the world of music in general and of black music in particular. The white Chicagoans sought out black cabaret music at an early age, but they generally did not grow up in neighborhoods that were rich in

African-American music. Rather, they lived in a separate white world that received its impressions of African-Americans through the media. Because the entertainment and recording industries were tightly segregated, Art Hodes, for example, had reached the age of twenty-three before even hearing a record of a black person playing jazz. He subsequently listened carefully to the sounds in South Side clubs. Max Kaminsky had played popular dance music professionally for six years before hearing the records of the black jazz pioneers.

The young white "Chicagoans" began their jazz journey by listening to the phonograph records of the leading white dance bands of their youth: Ted Lewis, Art Hickman, and Paul Biese. These bands remained a major, if often frustrating, influence on them. However sweetly corny or musically slapstick such dance music later seemed, it carried, when they first heard it, a thrill that they associated with the fox trot and up-to-date social dancing. It was their first escape from the waltz, violins, and dreamy, tearful, sweetly sentimental dance music. Eddie Condon, who made a career out of disparaging sweet bands and promoting hot ones, listed the recordings of dance band pioneer Art Hickman (and those of the Paul Biese Trio, Edison Trios, and Ted Lewis) among the first jazz records he heard.[26]

In their progress toward jazz, "the Chicagoans" soon developed a taste for records by the white Original Dixieland Jazz Band, who played a brand of music which they later repudiated as corny and out-of-date; but at the time they first heard it, the ODJB's sharp ragtime syncopations and sardonic, vulgar animal imitations had an appealing impudence. The smoother, more swinging beat of the New Orleans Rhythm Kings, and the King's serious musical attitude made a longer-lasting impression on them, in part because some listened to NORK in person at Friars Inn, as well as on records. This band, a mixture of New Orleans with midwestern musicians, dropped most of the clowning attitude of the Original Dixieland Jazz Band and took a more serious approach to jazz, even if they did become known for their boyish pranks on stage. As Ralph Berton, brother of Chicago percussionist Vic Berton and Bix Beiderbecke's biographer, put it:

> They played not in the zany, tongue-in-cheek spirit of the white bands ... but *seriously*—mean and low down, pretty or funky, driving or lyrical, but always *for real*. As we said in those days—there was no higher praise—*they played like niggers*.[27]

While still in high school, the Austin High Gang learned to play jazz by playing along with NORK records, so that their most immediate musical models were white. Jimmy McPartland described their apprenticeship:

> What we used to do was put the record on—one of the Rhythm
> Kings', naturally—play a few bars, and then all get our notes.
> We'd have to tune our instruments up to the record machine, to the
> pitch, and go ahead with a few notes. Then stop! A few more bars
> of the record, each guy would pick out his notes and boom! we
> would go on and play it. Two bars, or four bars, or eight—we
> would get in on each phrase and then play it all ... It was a
> funny way to learn, but in three or four weeks we could finally play
> one tune all the way through—*Farewell Blues*. Boy, that was our
> tune.[28]

McPartland's recollections also underscore the influence of pianist, composer, and band leader Elmer Schoebel, who, more than any other individual, charted the major musical course of white jazz in twenties Chicago. Born in 1896 in East St. Louis, Illinois, Schoebel was about ten years older than the rebel white jazzmen. He worked in popular entertainment from a very early age, playing piano in movie houses from age 14 and in vaudeville theaters beginning at age 16. Schoebel came to Chicago in 1919 with the 20th Century Jazz Band and went on to play a key role in creating the arrangements for the New Orleans Rhythm Kings. He designed a successful amalgamation of jazz and social dance music at the Midway Gardens in 1924–25, creating the hottest white dance band in a major Chicago ballroom, and went on to play and arrange for other top dance bands led by Isham Jones and Louis Panico. Most important, he wrote several of the tunes which came to define the literature of Chicago jazz, as opposed to New Orleans jazz. His "Farewell Blues," "Bugle Call Rag," "Nobody's Sweetheart Now," and "Prince of Wails" became important vehicles for the recorded improvisations of Chicago jazz musicians during the twenties and thereafter.[29]

Chronologically the last, and in some ways most powerful, white musical influence on the "Chicagoans" was cornetist/pianist Leon "Bix" Beiderbecke, particularly the records Beiderbecke made in 1924 with the Wolverines. If the New Orleans Rhythm Kings contributed a tough, white working-class spirit, Beiderbecke brought to the Chicagoans a mixture of exuberance and disciplined harmonic and tonal refinement.

The central cult figure of the Jazz Age, Bix was a middle-class Mid-westerner become ill-fated artist, lost in the big city; his dreamy abstraction had little in common with the rough-and-tumble inner-city street life, but he exerted an enormous musical influence on musicians filled with more jazz spirit than musical vision.

Beiderbecke actually spent relatively little time in Chicago. In November 1921, while still a teen-ager who had been banished by his worried parents to Lake Forest Academy, thirty-five miles northwest of Chicago, he got their permission to spend Thanksgiving in Chicago, and there he heard the New Orleans Rhythm Kings at Friars Inn. He thereafter snuck down to the Loop regularly during the semester, a practice which led to his expulsion from school in May 1922. When not listening carefully to the NORK at Friars Inn, Bix cultivated social and musical relations with Bill Grimm and a group of Northwestern University students who played jazz for fraternity parties and listened to the black cabaret bands on the South Side. In June 1922, Grimm got Bix a job playing on a Graham and Morton line excursion steamer on Lake Michigan. Beiderbecke developed a beautiful, unique tone on the cornet. When Eddie Condon heard him play that fall, he remarked that Bix's "sound came out like a girl saying yes."[30]

Beiderbecke and the Wolverines, a group largely composed of collegiate musicians from the greater Chicago area, played a brief engagement at Palmer Cady's Cascades Ballroom at Argyle Street and Sheridan Road on the Far North Side, jamming after hours at Dinty Moore's, a speakeasy at Broadway and Balmoral. They travelled to Richmond, Indiana, to make their influential records of 1924.

In the spring of 1925, Beiderbecke played at George Leiderman and Sam Rothschild's Rendez-Vous Cafe at West Diversey and Broadway on the North Side, a Gold Coast cabaret noted for "the pulchritude of its female ensembles" and for Charlie Straight's dance band. Although the club manager allegedly disapproved of Beiderbecke's modern harmonic ideas, Straight and his trumpeter Gene Cafarelli insisted that he remain, pleading that Bix was far more than just another cornet player. They even tithed themselves to make up a salary for him. Bix also joined the after-hours "relief" band which improvised for listeners, after Straight's orchestra had played ensemble dance music.[31]

And that, despite a brief notice in the summer 1927 *Orchestra World* that Beiderbecke was playing third trumpet in the Stevens Hotel Orchestra under Armin Hand and Roy Bargy, was the extent of Beiderbecke's

Chicago career. Even though *Variety* described the Wolverines as being "from around Chicago" when they made a hit at New York's Cinderella Ballroom in September 1924, they were more active on midwestern college campuses than in Chicago itself. And by 1925, Beiderbecke had accepted a job in Detroit with the Jean Goldkette Orchestra.[32]

Bix's records, however, inspired the white Chicagoans—particularly Max Kaminsky and Bill Davison—with his harmonic insight, improvisational poise, and restrained passion. He circulated on a collegiate social level which overlapped that of Oak Park and Austin, but the downtown, white Chicago sound was more fully captured by Jimmy McPartland, who replaced Beiderbecke in the Wolverines in October 1924. Despite Beiderbecke's deep influence on his style, McPartland played with a raw, nasal tone—the sound that Mezz Mezzrow called "a Bronx cheer for the righteous squares"—and a more impressionistic harmonic awareness.

Most of the white Chicagoans brought their white musical influences with them to their encounters with South Side music and culture. But a few had discovered African-American dance hall and cabaret music at an early age. George Wettling, who lived in Woodlawn, the all-white Far South Side rooming house district, had to travel through the black belt on his way to the Loop, and was just into his teens when he began stopping off at 31st and Cottage Grove. He was an early habitué of the Lincoln Gardens, where the drumming of Baby Dodds deeply impressed him. Muggsy Spanier was only eleven or twelve years old when he first sat outside of the Pekin Inn to listen to Joe Oliver's cornet playing. Spanier, who captured much of Oliver's approach with the cup mute, also listened to Oliver when he performed in Lawrence Duhé's band in the bleachers at Chicago White Sox home games.[33]

With the exception of Wettling and Spanier, most of the other white Chicagoans were in their later teens when they first went to the South Side clubs. Jimmy McPartland, Bud Freeman, and the Austin High Gang first heard about the South Side club scene when they played for fraternity parties at the University of Chicago and Northwestern. According to Freeman, Bill Grimm of the University of Chicago first took him to hear Oliver and Armstrong at the Lincoln Gardens. Onah Spencer, a South Sider himself and a writer on Negro Music in Chicago for the WPA Federal Writers' Project, named Murphy Podolsky, Fritz Neilson, Ted Clark, and Jack Kirk as instrumental in bringing the Chicagoans to the South Side for the first time. Condon identifies Podolsky as a pianist with "an inside track on school dance dates."[34]

Jazz acted as a shared sensibility and a musical craft which muted racial tensions between black and white musicians. This shared musical exploration gained its emotional impact from the powerful tensions of American racial attitudes. The young white Chicago jazzmen were powerfully drawn to the Lincoln Gardens. Their determination to invade this self-consciously African-American dance hall where King Oliver's band had been playing for three years, created tensions which surfaced immediately in their contacts with doormen.

Veiled remarks indirectly indicated racial attitudes to which neither blacks nor whites could refer directly. The young whites were kept waiting by the doorman until a series of signals indicated that they might enter. On their way in, the doorman often commented upon their reasons for coming. Bud Freeman adopted a strong southern black stage accent when remembering an extremely heavy doorman who always remarked: "Ah hears you-all's here to gets your music lesson tonight." Freeman chalked up the remark to the man's perceptiveness: "We were there for the music."35

South Side chronicler Dempsey Travis writes that Bill Summers, doorman at the Sunset Cafe, made similar remarks: "Good morning! I bet I know why you boys are back again this morning. You came for another music lesson, didn't you?" Such remarks, coming from the doorman of a black-and-tan designed to attract whites, probably were made with amiability, but they took on different meanings when placed within the context of the white musicians' professional interest in black music. One cannot forget that black musicians were not allowed to enter the Loop hotels and clubs in order to study the latest musical effects being produced there.

The casual manner in which white musicians sometimes replied to the comments of Bill Summers at the Sunset Club revealed a nettling insensitivity: "Some would reply, 'You're right, Professor.' Others would say, 'Here's a quarter, Sam. Get yourself a cigar.' The more timid ones only nodded, sometimes with a smile." Eddie Condon wrote insightfully that the young whites "had good reason to feel slightly uncomfortable until they had pushed their way close to the bandstand and been recognized by Oliver. A nod or a wave of his hand was all that was necessary; then the customers knew that the kids were all right."36

The actual physical movement from the sidewalk into the black dance hall—crossing a threshold that separated two distinct areas, was, as Victor Turner puts it, a rite of passage for the young whites. The black doorman's opening of the door for these whites, their subsequent crossing of a racial frontier, separated the young initiates, in their own minds, from everyday life and took them into a pulsating new realm of intense nonverbal experience that was beyond everyday routines. Edmond Souchon recalled passing through a dark winding hallway, feeling his excitement mount as he approached the wildly animated dance hall of the Royal Gardens Cafe. Once they had emerged into the brightly lit dance hall inside, the young white jazzmen felt that they had never before experienced such explosive excitement. Eddie Condon later recalled:

> As the door opened the trumpets, King and Louis, one or both, soared above everything else. The whole joint was rocking. Tables, chairs, walls, people, moved with the rhythm. It was dark, smoky, gin-smelling. People in the balcony leaned over and their drinks spilled on the customers below.... Oliver and Louis would roll on and on, piling up choruses, with the rhythm section building the beat until the whole thing got inside your head and blew your brains out.

Through an eager, heedless preoccupation with their own spiritual agendas, the whites, according to one black observer, often "literally muscled their way through the throngs of black dancers to get near the bandstands. Once they were there, they would hog that area until just before dawn." Condon and Freeman implied that the seating arrangements resulted from their status as musicians, that space near the bandstand was accorded them as a mark of respect. But at least one black observer claimed that the whites benefited from a Jim Crow seating policy.[37]

Their hypnotic fascination with the music created a realm of sacred time and space filled with intoxicating moments of emotional excitement in which young whites could defy lingering Victorian moral repression, racial segregation, and ubran-industrial discipline. The rebel Chicagoans recalled being completely hypnotized by the music. The Original New Orleans Creole Band with Sugar Johnny or Freddie Keppard on cornet, Sidney Bechet on saxes and clarinet, and Wellman Braud on string bass put Mezz Mezzrow "in a trance" and left him "breathless." "That was my big night," he wrote, "the night I really began to live." Eddie Condon recalled a similar flash of lucid understanding the first time he heard King Oliver's Creole Jazz Band:

It was hypnosis at first hearing. Everyone was playing what he
wanted to play and it was all mixed together as if someone had
planned it with a set of micrometer calipers; notes I had never
heard were peeling off the edges and dropping through the middle;
there was a tone from the trumpets [*sic*] like warm rain on a cold
day. Freeman, McPartland and I were immobilized; the music
poured into us like daylight running down a dark hole.[38]

Despite their extensive prior exposure to white bands and records, the
white "Chicagoans" were deeply moved by South Side music. Louis
Armstrong, of course, had a stunning impact on Wild Bill Davison,
Wingy Manone, and Max Kaminsky. Davison provides a particularly
vivid and clearly documented example: his earliest recordings with the
Cincinnati-based Chubb-Steinberg Orchestra reveal the influence of Bix
Beiderbecke's phrasing on his early playing style; once he heard Louis
Armstrong, however, Davison changed remarkably; the Beiderbecke in-
fluence all but disappeared, replaced by Davison's raging interpretation
of Armstrong's passionate flamboyance.[39]

Kaminsky's transformation followed similar lines. Already deeply
moved by Beiderbecke's tone and feeling, the white trumpeter never
heard Armstrong in person until January 1929, when he listened to him
in the Carroll Dickerson band at the Savoy Ballroom. "I felt as if I had
stared into the sun's eye. All I could think of doing was to run away and
hide till the blindness left me." Kaminsky explained what he heard in
Armstrong's playing:

Above all—above all the electrifying tone, the magnificence of his
ideas and the rightness of his harmonic sense, his superb technique,
his power and ease, his hotness and intensity, his complete mastery
of his horn—above all this, he had the swing. No one knew what
swing was till Louis came along.

Kaminsky expressed what most of the Chicagoans must have felt,—that
Armstrong brought grandeur to a form of music then associated with
song-and-dance entertainment and social rebellion.[40]

Drummer Baby Dodds exercised an equally profound influence over
Dave Tough and George Wettling. Both young whites heard him with
the Oliver band at the Lincoln Gardens, but also had ready access to
Dodds' music during his long tenure in his brother Johnny's band at Bert

Kelly's Stables in the Loop. Both emulated Dodd's equipment—the deep snare drum, the head loosely stretched, and Dodds' combination of cymbals and wood blocks—while imitating the South Side drummer's style of sensitive accompaniment and his use of fills and press rolls.[41]

Five of the white clarinetists—Benny Goodman, Rod Cless, Joe Marsala, Frank Teschemacher, and Mezz Mezzrow—took early inspiration from Jimmie Noone. Goodman also was influenced by clarinetists he heard on some of the early jazz records made by white New York studio bands, as James Lincoln Collier argues, but he specifically mentioned Jimmie Noone in his autobiography. Goodman, Noone, and Buster Bailey studied with the same teacher, Franz Schoepp; and Goodman, therefore, would have known Noone well enough to contact him at the Apex Club. Like Noone, Goodman played jazz with a full, rounded "legitimate" tone achieved by following the traditional European encouragement of full abdominal support, an open throat, arched palate, and lip—not tooth—support of the mouthpiece.

Goodman does not seem to have emulated other characteristics of Jimmie Noone's improvisational style—his regular eighth notes, staccato phrasing, and vertically oriented, broken chord clusters—but the white drummer and band leader Ben Pollack detected a transformation in Goodman's playing from his early fascination with vaudevillian Ted Lewis to "a mixture of Jimmie Noone, Leon Rappolo, Buster Bailey, and other great clarinet players." White clarinetist Joe Marsala, for whom "Noone was the man," confirms that Goodman "came around listening to Noone too." Moreover, Goodman was to popularize the type of small swing group with clarinet lead which Jimmie Noone introduced in Chicago. Goodman emulated much of that group sound in hiring pianist Teddy Wilson, who had taken inspiration from Noone's Apex Club pianist Earl Hines. Goodman also played the better known numbers from Noone's repertoire—"Four or Five Times," "My Baby Rocks Me, with One Steady Roll" (disguised as "Six Appeal"), and "I Know That You Know."[42]

Frank Teschemacher's dirty-toned, hell-for-leather, spikey improvisations also left an obvious impression on Goodman. Teschemacher also listened closely to Jimmie Noone, even if his own playing developed along quite different lines. His 1928 solo on "Darktown Strutters' Ball" with the Jungle Kings reveals a direct paraphrase of Noone's style: the young white reedman plays clusters of quarter note triplets followed by staccato eighth notes in the chalumeau register, concluding with an upward leaping sixteenth note broken chord capped by a dotted quarter accented with a wide vibrato.

White Iowa-born clarinetist Rod Cless, who played extensively in Chicago near the end of the decade, also credited Johnny Dodds and Jimmie Noone as major influences on his playing style.[43] But none of the white Chicagoans was more deeply or dramatically impressed with South Side musical culture than Mezz Mezzrow, who found in black music and musicians an originality, uncomplicated honesty, humor, and lack of perversity which, in his mind, contrasted sharply with the hypocrisy and racism of white society. He admired the mixture of lament with un-affected good spirits in South Side jazz, and the calculated grace of the black performers of his acquaintance; he used their lives and their art as repudiations of what he perceived as white middle-class racism and violence in America. Mezzrow adopted the mannerisms of black show business performers, an urbanized southern black accent, and black street slang. He married a black woman, and lived in Harlem for many years after leaving Chicago. He even claimed that his skin darkened as well. As one black observer put it, probably referring to Mezzrow:

> The Austin bunch were not just imitators, they absorbed the very spirit of this music ... and did more to introduce and set an example of real friendship between the races.[44]

Mezzrow went further than the rest, but South Side jazz also led Bud Freeman to a deeper appreciation of African-American grace under pressure:

> Now here were these black people who were allowed no privileges. They were not allowed to come into our shops and cinemas, but we whites were allowed to go out to their community, where they treated us beautifully. I found their way of life equally as important as their music. It was not just their music that moved me but the whole picture of an oppressed people who appeared to be much happier than we whites who had everything.[45]

This basic grasp of the African-American dilemma drew the white Chicagoans into a more sympathetic stance toward black jazz and deep-ened their sense of alienation from the white middle class. Some—like Dave Tough—returned to the South Side often, visiting not just the most prominent cabarets, but some of the smaller clubs, the rib joints and buffet flats about which most whites were ignorant. Tough ultimately married Casey Majors, an African-American cabaret dancer. Wild Bill Davison hung out in "an all-black place" called "the Ranch," and once he was established as Louis Armstrong's friend, he could go there without "anybody making any fancy remarks." Bud Jacobson attended all-black functions like the Okeh [Records] Cabaret and Style Show, one of a small number of whites in a crowd of thousands.[46]

In moving closer to South Side Chicago music and musicians, the young whites epitomized America's racial dilemma. South Sider Milt Hinton expressed the problem with telling irony:

> We didn't fraternize with white guys. White guys always knew
> where the good players were. That's the reason that Benny Good-
> man admits that he was tremendously influenced by Jimmie Noone
> in Chicago. Because he could come up to the Elrado, and the
> Golden Lily and the places where Jimmie Noone played and listen
> to him ... But we didn't go downtown to the College Inn to hear
> Ben Bernie, you see. It wasn't chic.

Wild Bill Davison put it bluntly: "They were always happy to see white people come ... because we had the money ... so there was no bad treatment. You got the best of it, really, whether they liked you or not."[47]

Davison and Freeman were among the few white Chicagoans to address directly the power advantage most white musicians enjoyed over black musicians under racial segregation. As whites, they were privileged visitors in the South Side clubs. A few pushed to sit in with the black bands, and, while acknowledging the compliment involved, black musicians wondered at their temerity. South Side reedman Scoville Brown, whose distinguished career got under way in the late twenties, recalled that the Austin High Gang were "still wet behind the ears ..." in the late twenties.[48] Nevertheless, according to Teschemacher-inspired clarinetist Bud Jacobson, Bud Freeman sat-in frequently on tenor saxophone with black bands, even though he was just beginning to learn. Freeman never wavered in his praise of black musicians, underlining in his many interviews, books, and autobiography their unquestioned priority in the invention of jazz. After hearing the Oliver band, he said, he completely abandoned his earlier white inspirations.

Jimmie Noone noted that, not content to just listen, the young Joe Sullivan, "not much more than a novice," rushed to sit down at the piano when the highly accomplished Earl Hines took a break. When Hines organized his crack show band for the Grand Terrace Cafe, white musicians, including Benny Goodman, Tommy Dorsey, Jimmy Dorsey, Muggsy Spanier, Joe Sullivan, Jess Stacy, Bix Beiderbecke, Hoagy Carmichael, and Frank Trumbauer, sat in. Hines remarked that he and his fellow black musicians "... sat around waiting to see if these guys were actually going to come up with something new or different."[49]

Black musicians allowed white musicians to sit in for several good reasons: first, the excitement of mutual discovery and the usually interesting and occasionally exciting joint explorations in musical improvisation created real bonds between black and white Chicago musicians in the twenties. As one black observer put it:

> The growth of hot jazz in Chicago during the twenties ...
> brought about the spectacle of white and colored "experts" meeting
> in order to further a common interest ...

Secondly, given the nature of American race relations, the whites' eager admiration of South Side music was flattering to most blacks. Dave Peyton of the *Defender* proudly noted that "many Ofay musicians" crowded Dreamland Cafe in January 1926 to hear Louis Armstrong "blast out those weird jazzy figures."[50]

Black musicians also allowed whites to sit in because the whites were at a different stage of musical development. They were usually younger, less advanced jazz improvisors who were not likely to outshine the black stars. Aspiring white jazzmen had few important club jobs playing jazz in Chicago, even at the height of the jazz age. Bud Freeman remarked that their repeated trips to the Local 10 union hall produced few results; he never held a long-term job with any band until 1933. Davison played with Benny Meroff's stage band, Teschemacher with Jan Garber's dance band, Spanier with Ted Lewis, and so on. As Art Hodes wrote: "We learned our jazz in Chicago, but we had to come to New York City to be able to pursue jazz as a livelihood."

Hoagy Carmichael remarked that Chicago's fledgling white jazzmen "just stood and waited. There were never enough jobs and too many new kids who thought the world was waiting for the sunrise of their talents. Chicago style came out of a lot of this standing around...." By comparison, black musicians like Jimmie Noone and Joe Oliver held down regular jobs that lasted for years, playing jazz in Chicago. The whites mostly played one-nighters, many of them on Northwestern, University of Chicago, and Indiana University campuses and summer jobs at lakeside resorts like Lake Delavan in Wisconsin. The black musicians could afford to humor them.[51]

The relative dearth of jazz jobs in white Chicago resulted from several interrelated factors. First, the hotels and dance halls were presenting dance music which the public accepted as jazz. Second, the major downtown clubs like Friars Inn featured floor shows, and several of the "Chicagoans" were not known for reading music. When the New Orleans Rhythm Kings broke up in 1923, Friars Inn continued its white New Orleans music policy and hired the Merritt Brunies Orchestra for a three-year residency, through the family connection between NORK trombonist George Brunies, and his trumpet-playing brother Merritt, whose "good standard all-around band" played a "competent" floor show and "good dance music." It was not until Brunies left for the Cinderella Ballroom in September 1926, that drummer Bill Paley put together some of Charlie Straight's ex-musicians with Jimmy McPartland, Jim Lannigan, and pianist Vic Breidis. The Austin High Gang was just moving into Friars Inn when the place was padlocked in June 1927 for violations of the Volstead Act.[52]

Although they played few long-term downtown jazz jobs, aspiring white musicians did take three-month jobs in summer resort dance halls like those at Delavan Lake, Wisconsin, Paw-Paw Lake, Michigan, and Hudson Lake, Indiana. Bix Beiderbecke, Eddie Condon, Art Hodes, Orville "Bud" Jacobsen, Dave Tough, Joe Sullivan, Murphy Podolsky, Wop Waller, George Wettling, Mel Stitzel, and many other young white jazz musicians spent the summers performing for vacationing college students and their parents, gaining invaluable playing time and enlarging their repertoires with a wide range of popular songs. During the daylight hours, they had plenty of time to listen to race records and develop their appreciation for the South Side musicians.[53]

Black musicians also permitted the young whites to sit in because several of the more serious of them—Muggsy Spanier, Frank Teschemacher, George Wettling, Joe Sullivan, Rod Cless—had begun to incorporate into their own playing elements of the styles of King Oliver, Jimmie Noone, Baby Dodds, Earl Hines, and Johnny Dodds. The depth of their artistic admiration encouraged black cooperation. Finally, there was always the chance that musical contacts with white musicians might lead to more jobs for black bands in white-only clubs and dance halls.

But black attitudes toward the white Chicagoans were often ambivalent. Some black musicians like Scoville Brown insisted that excellent relations developed between themselves and white jazzmen: "White and black musicians got along fine. Now, there were existing conditions in society that prevented our working together. You follow me? This is a social problem. But this isn't a problem among the musicians themselves."[54] But white appropriations of black jazz instrumental styles raised the specter of racial exploitation. In a racially segregated society, white musicians could qualify for lucrative playing jobs denied to blacks because of their skin color. The general popular ignorance of the South Side scene made it possible for white instrumentalists to entertain unsophisticated white audiences with techniques and ideas "borrowed" from blacks.

South Side musicians of the twenties referred to white musicians as "alligators," exploiters who copied black musical inventions in order to profit from them. Chicago jazz lore includes stories of Joe Oliver cutting the titles off the tops of his band charts and refusing to identify particular numbers in which white musicians took a suspiciously close interest. Dave Peyton questioned Oliver's encouragement of white jazzmen and became one of the first black writers to try to establish the priority of blacks in the invention of jazz.[55] Musicians usually referred to these problems more obliquely. Lil Hardin Armstrong, for example, remarked to Bud Freeman years after Chicago's heyday was over: "We used to look out at you all and say, 'What are they all staring at? Why are they all here?'"[56] Jelly Roll Morton said nothing directly, but his amanuensis, ethnomusicologist Alan Lomax, wrote in Morton's autobiography that the Chicago High School kids "did well; not only were they talented, they were the right color. [They] would have the money and the fame while old Doctor Jazz died hard in Atlanta and his boys were still scuffling in the honky tonks."[57]

When John Lax asked South Side drummer Red Saunders how close the black and the white musicians became in Chicago during the twenties, he replied: "Hm, *Comme ci, comme ça*—just so so." Ralph E. Brown recalled that musicians of both races drank together in South Side clubs after the whites had finished their night's work, but they were "not too close. We fraternized a little bit." Even the more optimistic Scoville Brown, when asked if Chicago in the 1920s had produced more closeness between the races, replied, emphatically, "No."[58]

The ambivalence of black Chicago musicians' attitudes toward the white jazzmen, colored by the history of American race relations and professional competition, was echoed in a similar pattern of ambiguity in white attitudes toward South Side musicians. The complexity of these perceptions was most dramatically expressed by Mezz Mezzrow, who articulated an extreme primitivist position which toppled into a kind of reverse racism of its own. Mezzrow's admiration of New Orleans jazz on the South Side emphasized its instinctual, guileless originality. He contrasted South Side jazz with white, east coast jazz which, he felt, relied solely on instrumental technique and legitimate tonal quality. The "true," "real" blues-rich New Orleans jazz, therefore, seemed to him to be innocent. He admitted that Oliver and Armstrong moved away from their "real" New Orleans music toward big band jazz as the twenties progressed, but he could not accept that such changes could represent a growing interest in new musical possibilities; to Mezzrow, it could only be commercialism.

Mezzrow's descriptions of his South Side experiences contain several such striking anomalies. He tells of a long evening in 1927 which he spent with the other white Chicagoans who were about to leave for New York, at the Nest, an elegant after-hours black-and-tan cabaret which advertised in the programs of various Loop theaters. At the Nest (actually called the Apex Club at the time) Mezzrow and his white jazz musician friends listened to Jimmie Noone's Orchestra, described by columnist Dave Peyton as playing "soft, scintillating, and sweet" music to packed houses. Mezzrow noted that Noone usually played "arrangements of the popular tunes of the day," but that, whenever Mezz and his white musician friends took their seats at the special table reserved for them in front of the bandstand, Noone would respond to their requests for "low down New Orleans gut bucket" blues. On this particular night, Noone agreed amiably, but, to Mezzrow's displeasure, actually played only show tunes. Finally, just as the other whites were about to leave, Noone and his saxophonist Joe Poston began to sing some blues lyrics which commented upon the tensions Mezzrow felt between himself and the other Chicagoans. Strikingly, Mezzrow interpreted the whole incident as reflective of his solidarity with "Negroes," and had nothing to say about the wealthy clientele to which the Apex Club catered, or about Noone and Earl Hines' steady refinement of their instrumental improvisations, or why he had chosen this band, made up of some of the more advanced jazz instrumentalists in Chicago, to ask them to play "low down New Orleans gut bucket" blues.[59]

Moreover, Mezzrow presented the most extreme version of a kind of well-intentioned, racist generalizing to which many of the young whites were prone. While observing black prisoners singing the blues, black jazzmen performing, and black vocalists and dancers entertaining in clubs frequented by the black sporting set, Mezzrow could not imagine either that the black performers were giving performances or that the sporting set was just one dimension of a far more complex, varied society on the South Side. Rather, he took what he saw and heard and generalized about "the Negro," who had a "simple and natural" bearing, a special "sense of time and rhythm that fascinated us," and a "relaxed, high-spirited, unburdened style of life." Mezzrow admired all of this focusing exclusively on those blacks closest to their southern folk roots and completely ignoring other black Chicago musicians like Sammy Stewart, Dave Peyton, Charles Elgar, and Erskine Tate who worked in the opposite direction, blending folk music with dance band arrangements and legitimately trained section men.

Finally, of course, Mezzrow forcefully articulated the major attitudes described in Norman Mailer's essay on the "white Negro." Mailer associated what he saw as a sociopsychological phenomenon with the beat generation after World War II, while making passing references to its roots in the America of the 1920s. According to Mailer, "hipsters" expressed their alienation from a war-torn, materialistic world in a mystic religion which focused on "The Negro," who appeared to offer a model for finding physical pleasure and musical expressiveness in the midst of despair.[60] Mezz Mezzrow was surely the most outspoken example of an historical phenomenon which stretched back through vaudeville and minstrelsy to the nineteenth century. As he put it:

> They were my kind of people. And I was going to learn their music and play it the rest of my days. I was going to be a musician, a Negro musician, hipping the world about the blues the way only Negroes can. I didn't know how the hell I was going to do it, but I was straight on what I had to do.

To his credit, Mezzrow, more than any of the other white musicians, at least raised the possibility that his own determination to play "Negro" music reflected elements of a white racist presumption that one could identify racial essences and come to possess whatever attributes of the oppressed race one might admire. He admitted that his discovery of the

South Side led him to think that it was "my own personal property." His collaborator, Bernard Wolfe, appended an insightful essay "The Ecstatic in Blackface" to a 1972 paperback edition of *Really the Blues*. In it, Wolfe systematically dismantled as "a coy fiction" Mezzrow's entire story of the spontaneous Negro, insisting, despite his own role as Mezzrow's amanuensis, that the true story was one of "the Negro as the white world sees him and forces him to behave."[61]

Cornetist Wild Bill Davison took a less romantic attitude toward the ambivalent relations between black and white jazz musicians. In 1926, Davison arrived in Chicago with the Seattle Harmony Kings and played at the Cinderella Ballroom, which catered to college kids out in Austin. He then caught on with Benny Meroff's highly successful vaudeville stage band, earning so much money playing in Chicago's leading downtown theaters that he was able to keep a table reserved every night at the Sunset Cafe. One night, Armstrong, who had never actually heard Davison play, invited him to sit in. Wild Bill played one or two passages which Armstrong might himself have played, and the black jazzman burst into laughter. Davison was never sure whether he was laughing with him or at him. Wild Bill later invited Armstrong to a party at his North Side apartment, where, it turned out, he was the only black. Armstrong refused to leave the kitchen. Davison and his friends therefore moved into the kitchen, too.[62]

With few clubs looking for genuine white jazz bands even during the winter cabaret season, the white Chicagoans who could read music moved into the dance bands. Those who couldn't read that well played in vaudeville acts like Red McKenzie's Mound City Blue Blowers, a novelty combination, or didn't work at all, dedicating themselves instead to "making the scene." Many made a habit of following Beiderbecke around from job to job, and Bix's arrival in Chicago with the Paul Whiteman Orchestra in 1927 led to one final, significant appropriation of African-American musical traditions by the Chicagoans—the "jam session."

A confirmed alcoholic from an early age, Bix sought out a place in which to drink in between the five shows a day that he was playing with the Whiteman band at the Chicago Theater. One block away, he discovered Sam Beers' My Cellar, "a blackened-up old store," at 222 North State Street, and began soaking up prohibition gin while exploring modern harmony at a battered upright piano in the basement. Soon this "fillmill," as Mezzrow called it, became an after-hours hangout for Chicago's white jazz musicians. Linking their music to images of sex and danger, they renamed the place "the Three Deuces, parodying the Four Deuces, one of the biggest syndicate whorehouses in town."[63]

Getting the leading soloists in the white big bands—Bix, Tesche-macher, Tommy and Jimmy Dorsey, Ben Pollack, Gene Krupa, and Benny Goodman among others—together with the rebel primitivists—Mezzrow, Condon, O'Brien, Red McKenzie, and Joe Sullivan—gave the whites another opportunity to learn the art of improvisation. And Bix was to be their teacher. Everyone wanted him to play (and to play beside him), not only because they admired his style but because they could profit from it. Mezzrow claimed that at the Three Deuces sessions, unlike informal sessions on the South Side, competition through solo "cutting contests" was rejected in favor of "really collective" improvisations in which musicians played together for pure enjoyment, after months of separation in different big and small bands. While that was probably true for some of them, others used the Three Deuces sessions for professional advancement.

Eddie Condon drank and played with the best men in the business, and got many of the musicians to come to the sessions. He met promoter and novelty musician Red McKenzie, and together they selected several of the musicians from the Three Deuces sessions for a recording date at Okeh Records. Bud Freeman, who was one of those chosen, was heard by drummer/band leader Ben Pollack, who hired him to go to New York. There, his career took on national scope.

Mezz Mezzrow would later claim that he had invented the "jam session" at the Three Deuces, giving these free-wheeling sessions their label by constantly requesting that they play Clarence Williams' "I Ain't Gonna Give Nobody None of this Jelly Roll."

> Down in that basement concert hall, somebody was always yelling over to me, "Hey, Jelly, what you gonna do?"—they gave me that nickname, or sometimes called me Roll ... and almost every time I'd cap them with, "Jelly's gonna jam some now," ... and I think the expression "jam session" grew up out of this playful yelling back and forth.

His claim receives little critical or scholarly confirmation, and the verb to "jam" and the noun "jam session" are credited to black musicians. But Mezzrow was an extreme example of the "embodiment" of African-

American culture among white Chicagoans. Thanks to racial segregation, he could "become," under special circumstances, Jelly Roll Morton.[64]

The rebel white Chicagoans, in their reach for something wilder, less saccharine, and less socially contrived than the music of the commercialized white dance bands, found themselves caught between the South Side jazz musicians, on the one hand, and a growing group of versatile white dance band professionals who played excellent jazz as well. The latter group—Benny Goodman, pianist arranger Mel Stitzel, Elmer Schoebel, Vic Berton, Frank Teschemacher, Volly de Faut, Muggsy Spanier, and Ben Pollack—assimilated the musical knowledge and sensibilities necessary to span the chasm between soulful expressiveness and dance band techniques, a perilous journey that was also made by many South Side jazz musicians. Some of the subsequently famous rebel Chicagoans like Eddie Condon remained proudly loyal to small group improvisation.

Taking an inflexible stand against the playing of scored arrangements of popular song materials lived to haunt Mezz Mezzrow, who later organized a historic inter-racial recording session in New York in 1933, mixing such black players as saxophonist and arranger Benny Carter, Teddy Wilson, bassist and bandleader John Kirby, drummer and band leader Chick Webb, and Chicago pianist/arranger Alex Hill with a few of his white Chicago colleagues. Hill later lamented that the arrangements he contributed to the session were ruined by the leader, who couldn't read them.[65]

White jazz spokesmen found it difficult to fully articulate a rationale for their desire to play jazz which did not rely upon race. Obviously something other than African-American music and culture provided their initial desire to play jazz, for few of the white Chicago jazzmen had heard black music before turning to jazz in the first place. The initial desire to play this tension-filled, fast moving music came from their anticipation of the excitement of urban life, their alienation from middle-class Victorian moralism, and a keen appreciation for the artful way South Side musicians played their music under trying cultural circumstances. Blacks, for a variety of specific historical reasons, took the lead in Chicago jazz, but the urge to play jazz music in twenties Chicago came from an interracial and polyethnic experience of early twentieth-century urban life.

5

Chicago's Jazz Records

THE PHONOGRAPH RECORDS of Chicago's jazz groups of the 1920s both responded to and acted upon the society that produced them. In this reciprocal process, economic and racial forces shaped recording and marketing decisions, while recorded jazz stirred a broader and deeper public appreciation of the talents of cabaret musicians. Of the two interwoven dimensions of jazz records, the music has received far more critical attention than the cultural or racial forces that shaped it. This chapter, therefore, will first examine selected dimensions of the record business as a context for the musical patterns engraved onto the records themselves.

The technology of sound recording revolutionized the public context in which jazz was experienced, lifting it out of the cabarets and inserting its sounds into homes across the country. Recorded sound technology proved an invaluable tool for ambitious jazz musicians, since records removed at least some of the distractions of the cabaret environment and encouraged record buyers to focus attention on the music. For the listener, repeated playing of particular phonograph records could reveal the musical excitement of jazz performance more clearly, a musical impact easily lost in live cabaret performance. Indeed, according to an article written in 1924 by Abel Green, an experienced entertainment reporter, few cafe customers in the twenties could even distinguish the melody of tunes they heard:

> Of 300 people in a cafe, about 40 percent are usually dazed and
> more inclined toward their respective vis à vis, the rest are dance
> mad and careless about the melodies, but stepping to anything with
> a rhythm, and, of course, there is the usual quota of "stoogies" who
> couldn't recognize the strains of the National Anthem in their con-
> dition ... 5 per cent are conscious enough to appreciate what it's all
> about.[1]

Eventually, jazz writers and musicologists appeared—from Hugues Panassié and Wilder Hobson to Gunther Schuller and Martin Williams—who relied nearly exclusively on phonograph records in writing
about the musical ingredients of jazz; both jazz's complex cultural func-
tion as cabaret entertainment and the economic and cultural influences
on jazz records were largely ignored.[2]

Some of the Chicago records by Oliver, Armstrong, Hines, Morton,
Beiderbecke, Noone, and Johnny Dodds have come to be interpreted as
milestones in the nearly seventy-five year history of recorded jazz. The
critically acclaimed jazz records made in Chicago during the twenties
have been selected from a larger number and broader range of records
made there during the decade. However, the subsequently famous jazz
musicians and jazz groups from that era also recorded many largely
forgotten dance numbers, vocal blues, and popular song performances,
and comic and novelty entertainment drawn from minstrelsy, medicine
shows, vaudeville, and cabarets. These latter sorts of records document
the musical enterprise of the decade, and, when they are regrouped for
purposes of analysis with the musically exceptional ones, the interactions
between music and cultural context emerge in sharper focus. The
broader range of Chicago jazz records reveals more clearly than the few
critically acclaimed ones how jazz functioned as a key ingredient in
urban entertainment enterprise. Jazz's musical evolution proceeded as
leaders, sidemen, and record producers worked to fashion dramatic and
entertaining musical acts with which to attract and hold public attention.

The musicians, jazz groups, and phonograph records that have been
most closely associated with jazz were only selected elements within a
musical culture generated by cabarets, theaters, and dance halls. Several
musical strategies, to be detailed later in this chapter, can be distinguished
within the larger body of jazz records made in twenties Chicago, different approaches which reflected an evolving artistic consciousness at work
under special conditions in a particular time and place. These musical

strategies responded to the economic and cultural pressures of that era, but they also were intended to draw public attention to instrumental music in cabaret, theater, and dance hall entertainment. In order more fully to appreciate the cultural context in which these strategies found expression, it will be necessary first to describe the evolution of jazz records within the record industry.

Jazz records first appeared as a specialized form of white dance music. Until 1913, little in the way of popular dance music was recorded in this country. American companies produced expensive phonographs and records for a wealthy elite. Operatic material dominated the recorded music, and phonograph manufacturers believed that customers bought records merely to have something with which to enjoy their phonographs. Beginning in 1913, however, the dance craze led to the production of dance records, which quickly turned immense profits and propelled the record industry forward during subsequent periods of economic weakness.[3] As early as 1916, record companies canvassed poor urban neighborhoods, like Chicago's South Side ghetto, to determine how many citizens owned phonographs. When enough inexpensive machines had been manufactured and sold—in the mid-twenties an Artophone suitcase portable costing $13.85 became popular on the South Side—records were designed to sell to specialized urban markets.[4] During World War I the number of phonographs rose remarkably; national production had reached $27,116,000 in 1914, but rose to $158,668,000 in 1919, representing 2,225,000 machines. Some companies even contributed new portable models to military units in order to help soldiers entertain themselves and develop a taste for recorded sound at their base camps during lulls in the fighting in Europe. The New York *Clipper* reported that growth continued through 1921, when one American out of every seventy-two owned a record player. In that same year, 100,000,000 records were made.[5]

In the dance music field, the crucial background against which jazz records emerged, the history of sound recording followed social and racial patterns established before the war: after the initial recordings of the white Original Dixieland Jazz Band were issued in 1917, jazz and dance records continued to be made overwhelmingly in New York City (and Camden, N.J.) by several companies using small white bands like the Ted Lewis groups, the Happy Six and other formations directed by Harry A. Yerkes, Earl Fuller's Famous Jazz Band, the Louisiana Five, Ladd's Black Aces, Bailey's Lucky Seven and other Sam Lanin groups, and the California Ramblers, as well as larger white hotel orchestras thought to appeal to middle-class customers and the carriage trade. During these years black band leaders like Wilbur Sweatman, James Reese Europe, W. C. Handy, and Ford Dabney recorded an interesting mixture of minstrel, ragtime, and orchestral music aimed at customers from both races.[6]

This white dominance, which ignored the long tradition of African-American musical entertainment of white audiences, seriously hindered black musicians. For both white and black musicians, making records formed an important ingredient in the planning and publicity that built performance reputations, particularly among those who had limited contacts with powerful white promoters and band leaders. Musicians could use their records to impress band leaders and band bookers, just as the latter used them to attract job offers from cabaret and vaudeville managers and dance hall proprietors. The appearance of the name of the cabaret or dance hall on the record label indicates that phonograph records were a form of advertising both within and outside of the music business.

Appearances on records also had a significant economic impact on jazz musicians. In the mid-twenties, whether or not their records ever sold in large numbers (the actual sales figures of all jazz records have been slow to surface), sidemen were paid $30 for each master that was cut and approved for production. When they accompanied vocalists, they could earn from $5 to $10, depending upon whether or not their instrumental work was featured during the recording. At a time when they often earned from $45 to $75 a week in dance halls and cabarets, recording fees could amount to more than wages.[7]

From 1917 to 1922, white jazzbands and the larger white hotel ballroom orchestras recorded prolifically in New York City. Chicago band leader Paul Biese traveled to New York as early as 1919 to record novelty numbers in the nut jazz tradition. The Isham Jones Orchestra was the only group to record in Chicago, itself, before 1922, waxing tunes like "Dance-O-Mania" as early as June 1920. The Edgar Benson Orchestra cut jazz-flavored records—"My Little Bimbo," "San," "Ain't We Got Fun?"—in Camden, New Jersey, beginning on September 20, 1920. Russo and Fiorito's Oriole Orchestra from the swank Edgewater Beach Hotel, and the Frank Westphal Orchestra soon followed the others eastward. Since jazz-influenced dance band records, like the Benson Orchestra's "Wabash Blues" (1921) and Jones's "Virginia Blues" (1922), appeared on the market first, the record-buying public, for a time at least, could have thought of Chicago jazz as something that the hotel dance bands often played. Lists in the Appendix indicate the band names, the chronological order in which they were recorded (expressed through the date of their first and last recording sessions of the decade), the number of jazz records they made, and the company labels involved.[8]

The subsequently celebrated records of black Chicago jazz bands were first made in 1923, six years after those by the Original Dixieland Jazz Band and three years after the first of Chicago's white jazz age dance records by the Paul Biese Trio and the Charlie Straight Trio. Thus they marked the systematic introduction of black groups into a field already dominated by a small number of highly productive white groups who recorded in New York City. In part, the decisions to make such records resulted from new, mass marketing strategies which involved, as *Variety* put it, "exploiting local bands for local trade, figuring on the band's domestic following and possible radio popularity to augment local trade."[9]

Decisions to produce black jazz band music on records intended for sale in African-American neighborhoods followed earlier efforts to profit from the sale of Irish, Yiddish, German, French, Native American, Hawaiian, Mexican, Bohemian, Polish, Tyrolean, and Scandinavian musics wherever these immigrant groups clustered. Such "Race Records" were intended to develop a working-class, urban, ethnic market for inexpensive phonograph records. As we have seen after World War I, black journalists referred to African-Americans as "the Race," so that the choice of the term Race Records did not necessarily carry negative implications for those meant to buy them. Much depended, however, on who used the term and the context involved, since within at least one record company (discussed below) that produced these products, racial market operations were conducted in a strictly segregated manner.[10]

One of the first companies to develop the urban black working-class market was the Okeh Record Corporation. In New York, on February 14, 1920, Okeh recorded black vocalist Mamie Smith singing "That Thing Called Love" and "You Can't Keep a Good Man Down." Sales of this first record were moderate, but Smith's subsequent August 10, 1920, recording of black composer Perry Bradford's "Crazy Blues" and "It's Right Here for You," both accompanied by her Jazz Hounds, sold 75,000 copies in Harlem alone over several weeks, demonstrating the existence of a lucrative market for popular black vocal music. In 1921 through

1922, a major craze for black female blues singers, whose work strongly reflected the world of urban show business, received encouragement from the Okeh Record Company, Paramount Records, and Vocalion Record Company. Although vocalist Mamie Smith's Jazz Hounds did make some important instrumental records in New York in 1922, and although many of the most influential black jazz instrumentalists played accompaniment on these "Classic Blues" vocal records, the concerted move to record instrumental music by black jazz bands did not begin until 1923.[11]

Chicago's Race Records subsequently reproduced several different kinds of sounds from a broad range of styles and tastes within the African-American ghetto community. In addition to the so-called Classic Blues vocal records by black female cabaret and theater singers, instrumental jazz records paralleled and competed with recordings by ragtime vocalists, minstrel, medicine show, and street corner entertainers, comedians and parodists, gospel and church choirs, jubilee and gospel quartets, jug and washboard bands, string bands, harmonica players, and an impressive number of recorded sermons.[12] These conceptually different sorts of recorded sound, moreover, influenced one another. Thus, what have been called "jazz records" often reflected elements of several forms of entertainment while at the same time setting standards for more purely instrumental musical effects in the dance band field.

Although precise sales figures are scanty, these race records apparently sold briskly. The long list of blues vocalists and jazzbands produced on record, after Mamie Smith's breakthrough, indicates a strong demand for racially oriented music. According to *Variety,* in mid-1923, "colored singers and playing artists are riding to fame and fortune with the current popular demand for 'blues' disk recordings." Clarence Williams, who had moved from Chicago to New York, told the New York *Clipper* that racial discrimination had created a demand for records among entertainment-hungry blacks, who brought records home for parties, since live performances occurred only in prohibitively expensive or racially exclusive clubs.[13] When the blues craze first hit, Williams had three music stores in Chicago on State Street—to handle record sales. Erskine Tate sold "Crazy Blues" in large numbers from his store at 31st and Wabash, and the *Defender* regularly advertised many other music stores that stocked race records. Harry Rife, proprietor of the Metropolitan Music Store at 47th and South Park, sold unprecedented quantities of records like Clarence Wiliams's "Gulf Coast Blues" and Bessie

Smith's "Back Water Blues." During the height of the rush, some customers paid two dollars to reserve copies of popular records which normally sold at half that amount or less.[14]

Race records also sold by mail. Newspaper advertisements for Paramount and Vocalion Records carried a small form which readers filled out, clipped, and sent to the company in order to receive (at a list price of 75 cents per record plus a cash-on-delivery charge) the products advertised. Moreover, door-to-door sales were more important than has been realized: race record advertisements included a call for full- or part-time retainers; the newsboys of the Chicago *Tribune* and Chicago *Defender* regularly carried copies of the latest records of the week along with their newspapers. They sold the disks at $1 apiece; for many customers the records were as important as the news. As one newsboy recalled:

> You'd go to one customer and she'd get all excited over a new blues and start in to telling you all about her girl friend or some relative who was sure to buy one, too.

Even Pullman porters augmented race record sales by carrying batches of the latest issues with them on trips from Chicago into the South, where country folk often enjoyed listening to the sounds of the city.[15]

Race pride and musical entrepreneurship mingled in South Side Chicago's attitudes toward the race record business. When New York's James Reese Europe cut "Too Much Mustard" and "Down Home Rag" on December 29, 1913, the first records made by a Negro orchestra, the Chicago *Defender* concluded that Europe was "Jazzing Away Prejudice," since whites usually encountered blacks only as porters, cooks, and waiters.[16] Moreover, when Mamie Smith's "That Thing Called Love" later appeared on the market, the *Defender* announced that "lovers of music everywhere and those who desire to help in any advance of the Race should be sure to buy this record."[17] Each subsequent effort by Okeh to drum up business in Chicago was met with further journalistic encouragement. The Okeh "Colored Folder," the first of the race record catalogues of titles by "colored artists" was greeted with applause; in response to lowered prices during the company's May 7, 1923, "Okeh Week" the *Defender* announced that "Racial Pride demands that full advantage be taken of this generous offer."[18]

Okeh's "Twelve Room House for Blues" campaign in November 1924 further revealed the mixture of racial pride and entrepreneurship that animated the race record business. Taking their cue from what their newspaper advertisement called "the splendid reception given by lovers of high-class music to the album sets of imported Odeon recordings," Okeh issued a hard-cover jazz record album: the front cover's design featured "a weirdly-tilted 'House of Blues' and its human overflow of laughing, dancing, 'blues' bands"; the pockets ("rooms"), into which record buyers would insert their "blues" disks ("tenants"), were framed by interviews with Clarence Williams, a major black jazz entrepreneur, blues artist Sippie Wallace, and vocalist Sara Martin. It was an "indirect appeal to the growing pride of Race."[19]

The Consolidated Talking Machine Company's promotional cooperation with South Side musicians and music entrepreneurs climaxed in 1926 with two star-studded programs staged at the Chicago Coliseum. The first, called "Okeh Race Record Artists' Night," took place on February 27, 1926, co-sponsored by the South Side Elks Lodge. It was prepared for by an intensive publicity campaign in the sixteen Okeh record retail outlets on the South Side. At the Coliseum, guitar sensation Lonnie Johnson, in company with Louis Armstrong's Hot Five, made records on stage "to demonstrate how its done." These recordings were played back to an awed crowd immediately afterward. Clarence Williams, Bennie Moten, King Oliver, and Richard M. Jones further entertained the crowd, leading orchestras which accompanied several heavily promoted vocalists. The program was broadcast over the Chicago *Tribune*'s radio station WGN, and ex-Mayor Thompson took the occasion to make a speech.[20]

Consolidated Talking Machine Company, the record distribution subsidiary of Okeh, also promoted the culminating event of the jazz movement in twenties Chicago: the "Okeh Cabaret and Style Show" organized by company president E. A. Fearn. Designed to promote Okeh stars and sales while also benefiting the black Local 208 of the musicians' union (which held parades for three days advertising the event), the show was held in the Chicago Coliseum on Saturday night, June 12, 1926. Okeh dealers sold tickets at discount with record purchases. Ten bands played, including the dance orchestras of Sammy Stewart, Erskine Tate, and Charles Elgar, along with two blues vocalists, and the comedy team of Butterbeans and Suzie. Louis Armstrong's Hot Five broke up the show.[21]

While intended primarily for black audiences, blues and jazz race records also attracted white customers from the start; given the sorts of black-and-tan cabarets in which most "classic blues" singers and jazz musicians worked, it would have been surprising if at least some whites had not bought these records. *Variety* routinely reviewed selected blues and jazz records by black musicians and reported that Okeh race records had "caught on with Caucasians," who accepted "the recognized fact that only a Negro can do justice to the native indigo ditties." Reviewing "Birmingham Blues" and "Wicked Blues" by Edith Wilson and Her Jazz Hounds, the trade paper decided that "the fair Caucasian percentage that dote on barbaric wails of the indigo order" had a good buy await-ing.[22] The New York *Clipper* highly recommended King Oliver records as "barbaric indigo dance tunes played with a gusto and much ado which leaves little doubt as to their African origin."[23]

In Chicago, Jack Kapp, creator of Brunswick Records' Vocalion race line and later Director of Decca Records, owned a record store at 2308 West Madison Street on the edge of the black ghetto that was emerging on the West Side. He stocked race records for white, as well as black, customers. Similarly, Tom Brown (leader of the Six Brown Brothers saxophone act) sold race records in his music store on South Wabash Avenue.[24]

White record companies fought bitterly to retain their monopoly on the recording and marketing of African-American music when Harry H. Pace organized his all-black Black Swan Record Company. Whites owned all but three of the race record companies and of those three, Sunshine Records, C & S Records, and Pace and Black Swan, only Black Swan made a major contribution to race record production. Even though black musicians recorded music primarily intended for black audiences, race records ultimately reflected the policies and ideas of the whites who owned the companies. White executives worked through black inter-mediaries, talent scouts and studio managers such as Clarence Williams in New York and Mayo Williams and Richard Myknee Jones in Chicago, who exercised substantial influence in guiding the development of black recorded music.[25]

Mayo Williams, who took charge of the race recording program for Paramount Records from mid-1923 through mid-1927, produced many excellent jazz records by Freddie Keppard, Johnny Dodds, and Tiny Parham in addition to his better known blues disks. His activities illus-trate how race records came to be made and in what different senses they represented African-American culture. Williams, the first black to hold an executive position in a white recording company, came to Chicago in

1921 and first wrote sports articles for the Chicago *Whip* before selling his services to the all-white Paramount Record Company, a faltering corporation which had purchased the catalogue of Black Swan Records. Williams represented himself to Paramount's sales manager M. A. Supper as having an inside knowledge of the black community's musical tastes, even though his own ran to opera rather than the blues and jazz. Eager to make money on the race market without having to become directly involved with the racial community, Paramount agreed to give Williams free rein and to pay him one cent on each record sold; he received no salary, but rather what he preferred to call "a talent" or "sales royalty" from each of the artists whose sessions he produced.[26]

Under Mayo Williams, race records came to be strongly influenced by South Side cabarets and vaudeville theaters. Since he had limited personal taste for popular black music, he scouted talent at the "new" Monogram Theater, the popular vaudeville theater on South State Street, two blocks from his own office in the Overton Bank Building at 36th and South State. As he was quoted as saying, "Nothin' but lowly people went to the Monogram ... the uppercrust went to the Grand." Williams stood backstage in the theater, measuring the applause accorded various vocal and instrumental acts. He followed the theatrical columns in the *Whip* and the *Defender* and frequented important black cabarets like Bottoms's Dreamland and the DeLuxe. Finally, Williams hired such important composer/pianist/arranger/accompanists as Tiny Parham, Thomas A. Dorsey, and Lovie Austin to identify musically interesting performers, teach proper pitch and arrangements to folk vocalists and musicians, and transcribe their subsequent performances for copyright lead sheets and sheet music.[27]

Paramount race records therefore both reflected and shaped African-American musical traditions. If whites owned the companies, black artists and producers exercised substantial control over the music chosen and recorded. The music had normally proven its popularity with South Side audiences before getting on record. Artists whose initial record did not sell at least 250 copies were not invited back to make more.

Mayo Williams personally regarded the separate retailing category printed in company catalogues as demeaning to blacks. Moreover, the company refused to allow Williams "to be identified with white records, or the white side of the situation at all." He was completely isolated from his employers, who never set foot in his South Side office, and when he was summoned to the yearly meeting with company officers in the Palmer House, he took the service elevator to the appropriate floor.[28]

The racially segregated organization of record companies encouraged certain types of race records. Black vocalists and musicians were encouraged to record what the white company executives and their black talent scouts considered to be distinctively "Negro" material. Such pressures encouraged racial stereotypes. Jazzman Danny Barker complained that black vocalists were forced by black record executives like Mayo Williams to do "earthy" material. Mamie Smith's record sales dropped when she started to record ballads instead of the blues that were expected of her. Instrumental music avoided the overt stereotypes of race and gender that attached themselves to vocal performances.[29] Jazz could not, however, entirely escape the economic and racial pressures of the record business.

Due in part at least to the heavier expenses involved in recording a ten-to-twelve piece orchestra, all of the record companies seriously under-recorded Chicago's large black theater and dance orchestras, which were, nevertheless, firmly entrenched in South Side movie theaters and a few leading white dance halls, and who should have been recorded far more than they were.[30] Racial stereotypes combined with economic pressures to encourage a preference among recording executives for five-to-seven-piece jazz bands, rather than the larger and more legitimate sounding orchestras. As Dave Peyton put it in his weekly column "The Music Bunch" near the end of the decade:

> In the past the big recording companies have confined our musicians to one style of recording. This style of recording they consider our orchestras are perfected in. They confine us to low jazz and blues. They have an idea that our orchestras cannot play real music for recordings, but they never were so wrong.[31]

Peyton was a theater pit orchestra leader and thus in a good position to understand how jazz polyphonies could act as a racial trap for ambitious musicians.

Harry Pace, unlike most record producers, took care to record a broad variety of musical genres when he founded the African-American record company called Black Swan, but his enterprise was short-lived.[32] Once created, stylistic categories helped to determine subsequent public expectations. As the twenties spun on, one trade paper indicated that,

despite the efforts of a handful of white musicians, blacks most authen-
tically recorded black music. The white writer wondered aloud at the
"paradox for one of his race" when Fletcher Henderson, recording with a
big band, "delivers a white man's blues style that is not at all faithful,
coming as it does from a crack negro aggregation." The writer felt that
whites had "done their damnedest to simulate the native negro 'blues'
and succeed indifferently with but occasional exceptions."[33]

Despite cultural and economic pressures on race recording, the move
to record Chicago jazz bands, which followed the decline of the Classic
Blues craze, helped to move cabaret musicians out from behind the
vocalists and other stage entertainers, spotlighting a form of small group
musical entertainment that avoided the more overt racial stereotypes of
cabaret show business. In 1922, Gennett Records, which had pioneered
mail order and chain store record distribution, began recording the latest
small group instrumental sounds from the city. Gennett, a subdivision of
the Starr Piano Company of Richmond, Indiana, developed the urban
ethnic and racial markets. It eventually sold its wares under inexpensive
labels like Champion, Buddy, Bell, Black Patti, Herwin, QRS, Challenge,
Conqueror, Superior, Supertone, and Silvertone. Following the established racial policies of the larger and more expensive labels like
Brunswick and Victor, Gennett's first efforts at small band instrumental
jazz had been limited to white New York-based groups like Jimmy
Durante and the Original New Orleans Jazz Band, Bailey's Lucky Seven,
and Ladd's Black Aces (whites drawn as caricatured Negroes for company advertising).[34]

The record industry's preference for white jazz bands continued in
Gennett's selection of the peppy jazz age social dance music of white
Chicago dance band promoter Husk O'Hare's Super Orchestra, recorded
at its Indiana plant on March 9–10, 1922. On August 29, 1922, the label
followed with records of the white Friars Society Orchestra. Its members
traveled from their regular engagement at Mike Fritzel's Friars Inn in
Chicago to Richmond, Indiana, to make the first recordings of Chicago
jazz band music, seven months before any black jazz bands did.

The choice of the Friars Society Orchestra for these historic records
followed directly from the group's popularity in white Chicago. The
Starr Music Company store was just around the corner from Friars Inn,
where the size of the crowds convinced Starr's Chicago manager Fred
Wiggens that a market for records by a small cabaret jazz band might
exist. For subsequent sessions at Gennett in the spring and summer of

1923, the group, which had ended its seventeen-month residence at Friars Inn, dropped its association with both O'Hare and the cabaret, taking their own name, the New Orleans Rhythm Kings.[35]

The impact of these early white jazz band records in Chicago is best understood against the background of instrumental novelty and dance records already made from 1917 to 1922. Compared with the nationally popular 1917 records of the Original Dixieland Jazz Band, the Friars Society Orchestra played New Orleans style polyphonic music seriously, very largely rejecting the "nut" humor of the earlier white group. Compared with the jazz age dance records already made by the Paul Biese, Husk O'Hare, Frank Westphal, Russo and Fiorito, and Isham Jones hotel orchestras, those by the Rhythm kings challenged the precisely executed, legitimate-toned music of professional dance bands with an earthier-sounding, coarse-timbered polyphony mixed with head arrangements devised by pianist Elmer Schoebel. Schoebel's own "Farewell Blues" and "Bugle Call Blues" and the band's performances of "Eccentric" and "Discontented Blues" all offered a judicious mixture of harmonized effects with polyphony. The band was well rehearsed, having performed their material repeatedly at Friars Inn. Schoebel, who could read and write music, taught his arranged effects to his young, non-reading musicians.

In these "head arrangements" individual parts were worked out by adjusting intuition and spontaneity with memorization of preconceived note patterns. With such arrangements, the records made by the Friars Society Orchestra, like those of several other important Chicago jazz bands, were frequently less fully improvised than the legend of jazz improvisation would suggest.[36]

The rugged individualism of the group's wind instrumentalists lent an earthiness to their sound seldom found among white bands. The guilelessly emphatic lead horn of Paul Mares, whose coarse-toned melodic statements rarely strayed from the middle register, dominated the front line sound. Just as importantly, the Friars/Rhythm Kings achieved a remarkably relaxed, rhythmically swinging beat and avoided the tendency of many white dance bands to sound nervous when playing "jazz." At the time they recorded, few popular white groups had ever achieved this flowing and focused rhythmic impact. Unlike many of the white hotel and dance hall bands of that time, in which a rhythmic disjunction separated percussion from wind instruments, the Friars Society Orchestra played with a shared rhythmic pulse.

But their influence on Chicago listeners did not increase with time. Groups of white musicians continued to assemble for NORK recording sessions, but with many changes in personnel; the core musicians in this group never worked again as a cabaret band after their initial run at Friars Inn and therefore lost much of the cohesion and musical focus which regular live performance with fixed personnel had encouraged earlier. Moreover, the subsequent move to record black jazz bands on "race records" soon revealed the depth and variety of regularly employed South Side jazz groups which had awaited exposure on record.

Indeed, in its own way NORK had helped to direct attention to the rich possibilities in recording South Side music when they recorded with Jelly Roll Morton at the piano on July 17 and 18, 1923, the first interracial recording sessions by Chicago musicians. But in addition to making an integrationist statement, NORK's record's benefited musically from Morton's loping left-hand swing that added even more propulsion to a good rhythm section; his beautiful right-hand countermelodies behind the soloists added a gracefully contoured counterpoint to their improvisations. On the other hand, the band's recordings of "Sweet Lovin' Man," "Wolverine Blues," and "Mr. Jelly Lord" gained in popular impact from the fact that the composers involved were African-Americans who had not yet been given the occasion to record them themselves. NORK's "Tin Roof Blues," their most influential number, long struggled against accusations of having been stolen from Richard Myknee Jones's "Jazzin' Babies Blues." What could be interpreted as unusually close relations with black jazzmen, might also represent the taking of unfair advantage.[37]

The marketing of race records swiftly uncovered the number, variety, and depth of jazz bands which had been hidden by the preference for white groups among the record companies. Although Clarence Jones made some solo piano records in Chicago in January, King Oliver's Creole Jazz Band made the first recordings of a black Chicago jazz band on April 6, 1923, seven months after the white Friars Society Orchestra, when Gennett, as part of a campaign to enlarge its race record catalogue, recorded nine numbers by them in one session. (Drummer Baby Dodds later explained that "we did all that recording in one day because none of us had quarters to sleep in Richmond.")[38] Fred Gennett had been introduced to Oliver by his Chicago store manager Fred Wiggens and, sensing sales possibilities in the large crowds which crammed the Lincoln Gardens, signed him to a recording contract immediately. The Richmond, Indiana, company therefore deserves the credit for giving Chicago's black jazz bands their first break into the recording business. Joseph Oliver recorded sixty-six sides in the Midwest with Gennett, Okeh, Columbia, and Paramount; only South Siders Louis Armstrong, Jelly Roll Morton, and Jimmie Noone were to become as prolific. Oliver made more records than any Chicago jazz band leader of the twenties except Jelly Roll Morton.

The Okeh Record Company soon set the pace for prolific recording and aggressive publicity for its race records, setting up a temporary recording studio in the Consolidated Talking Machine Company Building at 227 West Washington Street. The details of how Oliver and his band first came to record for the Okeh label confirm the importance of musical entrepreneurship on both the black and the white sides of the music business.[39] In early June 1923, a Music Trades Convention met in the Drake Hotel and presented a series of bands to entertain the conventioneers. King Oliver's Jazz Band was scheduled last on the bill and appeared at 3:00 a.m. for a twenty-minute set, which eventually ran nearly two hours longer than planned. What the journal *Talking Machine World* called a "mob of admirers" crowded around to marvel over the band, "especially the little frog-mouthed boy [Louis Armstrong?] who played cornet." The band recorded for Okeh later that month.

Important recordings of African-American music resulted from commercial decisions made by white companies to use black bands and to rely upon sales to Chicago's black population, black tourists, and a substantial number of whites who were thought to like black jazzband music. In September 1923, the General Phonograph Company took advantage of a convention of the African-American Improved Benevolent Order of Elks to advertise the Oliver and Erskine Tate bands' new records by tacking up posters and having the bands perform throughout the South Side on flatbed trucks.[40]

The records of Joe Oliver's Jazz Band revealed that polyphonic performance benefited from unusually distinctive and powerful instrumental voices. Like NORK, despite their breaks and interesting head arrangements, most of the Oliver Jazz Band records (more than those he later made under the name of King Oliver's Dixie Syncopators) focused on polyphonic performance: the cornet(s) carried the melody, the clarinet embroidered, and the trombone punctuated rhythmically. But Oliver's band, like most of the recording white dance orchestras of the time, had two cornets whose close harmonic voicing provided an unusually powerful lead, and their clarinetist Johnny Dodds played with unprecedented emotional power and depth. Moreover, Oliver's jazz band had discovered some challenging ways of arranging oral musical traditions so as to introduce a variety of sounds which enriched the overriding polyphonic paradigm.

Recordings of King Oliver's group demonstrate a mixture of improvisation with arrangements developed in oral/aural tradition music. Like the earliest NORK recordings, Oliver's involved, fairly set routines derived from rehearsals and four years of on-the-job experience. Indeed, some of their routines must have been New Orleans aural/oral traditions transplanted to Chicago. As noted earlier, Eddie Condon, thought that the mixture of individual instrumental voices in the Oliver Jazz Band sounded "as if someone had planned it with a set of micrometer calipers."[41]

The band recorded "Dippermouth Blues" on two occasions (two and a half months apart) in 1923. The performances are very similar, not only in the ensemble arrangements but also in the now-famous solo lines played by Oliver (an incendiary cup-muted wail) and clarinetist Johnny Dodds, who began his solo with three measures of keening whole notes. Indeed, when, three years later, Oliver recorded the number a third time (calling it "Sugar Foot Stomp") the cornet and clarinet solos still hinged on the same motifs. Similarly, two recordings of "Snake Rag," with its slithering, muted cornet-to-sliding-trombone effects, unfold in a strongly similar manner.[42]

Oliver wove a variety of different textures, requiring planning and memory, into his ensembles. First, even when they were polyphonic, the two-cornets-in-harmony lead added a widely admired element of preparation. Moreover, Oliver sometimes turned the lead over to his powerful clarinetist Johnny Dodds, supporting him with choir-like harmonies from the other wind instruments. He also varied the relationships between lead and supporting instruments, giving the lead, in the trio of "High Society," for example, to Dodds, playing in the chalumeau register, while Armstrong played hushed harmonies one octave higher. Armstrong's harmonic support, moreover, surpassed the simple statement of parallel chordal tones, forming instead a contrapuntal melodic line, key notes of which harmonized with important harmonic moments in Dodd's direct statement of the tune's trio theme. To make his band sound even more surprising, Oliver would change from polyphony to homophony and back again from chorus to chorus and within choruses.

Oliver carefully selected his repertoire in order to bring out of his musicians a variety of moods, extending well beyond the expected jazz age social dance excitement. In fact, one outstanding quality in the Oliver sound is the recurrent sense of other, sometimes sacred, emotions coloring what functioned as dance hall music. Johnny Dodds and Louis Armstrong brought a prayerful humility to the trio of "High Society" and to the moving "Riverside Blues," written by gospel composer Thomas A. Dorsey. So too, the stirringly powerful, repetitive melodic themes of "Working Man Blues," with their beautiful harmonies, gave Oliver's records a singular emotional power.

But the changing focus of the music found on records made by other, later groups led by King Oliver demonstrates that a band's performance situation in Chicago influenced the kinds of records it made. The Vocalion records which Oliver made with his Dixie Syncopators beginning in 1926, when the group worked as the house band at the Plantation Cafe, included musical and show business ingredients that were missing from his Royal Gardens Cafe dance band.[43] Oliver now included "legitimate" section men in his ten-piece Plantation Cafe and Vocalion recording band. Bob Shoffner, who went on to play with Tate, Elgar, and Peyton, read the scored cornet parts on "Too Bad" and "Deep Henderson" in 1926, while Albert Nicholas, Billy Paige, and Barney Bigard combined in a harmonically and tonally rich red section that was entirely absent from Oliver's earlier recording group. This saxophone choir, which caused the band's sound to more closely approximate that of most white dance bands, was particularly adept at changing tonal colors through varying combinations of the soprano, alto, and tenor saxophones with clarinets.

On some of his Dixie Syncopators records cut in Chicago, Oliver demonstrated an increased awareness and control of techniques implied on some of his Creole Jazz Band sides. Where clarinetist Johnny Dodds had sometimes played lead within polyphonic ensembles, Nicholas or Bigard (and even Johnny Dodds himself, who returned to the Oliver band to perform on "Someday Sweetheart") now stepped out of the ensemble to spotlight the melody and the clarinet sound, while the brass, or reeds (or sometimes both) chorded together in the background or punctuated at rhythmic intervals. This increasing differentiation between the improvising or reciting soloist and background accompaniment reflected a growing solo instrumental specialization and an orchestral stylization appropriate to cabaret entertainment.

The Plantation Cafe also hired singers, and some of the King Oliver Dixie Syncopators recordings therefore presented several different female cabaret vocalists. Oliver and his band backed Teddy Peters on "Georgia Man," a cabaret song, Georgia Taylor on the vaudeville-influenced "Jackass Blues" (composer and vaudevillian Perry Bradford in live performances of this tune, appeared on stage with a mule), and Irene Scruggs on "Home Town Blues" and "Sorrow Valley Blues" in the classic blues mold. None of these vocalists could match the musical talent of their accompanists. They probably came across better in live performance, where they could dance as well as sing. Each of them sang straightforward blues with lyrics that dealt with strife between the sexes, indulging in none of the cheaper entertainment tricks or sexual double entendres often found on other records.

Jazz records helped to sharpen a new focus on instrumental musical entertainment. In the competitive world of Chicago musicians, each jazz instrumentalist and each jazz band worked to develop a distinctive sound. Different combinations of musicians inevitably produced different group sounds, even when everything else remained much the same. Many interesting records by both white and black groups, for example, relied more or less completely on the basic jazzband paradigm of five to seven musicians playing polyphonic choruses on a blues or popular song. To what degree such ensemble work really was improvised is open to question. While these performances certainly do not sound as if they were being read from written scores, the musicians may have memorized and then repeatedly played the patterns of notes they recorded. The proponents of this style, as far as one can tell from their records, largely resisted the interesting head arrangements found in the early recorded work of NORK and Joe Oliver. Within it, one can locate Freddie Keppard's Cardinals, the New Orleans Boot Blacks and the New Orleans Wanderers (the same group under different names), Jelly Roll Morton's pre-1926 recordings, the State Street Ramblers, and Lovie Austin and Her Blues Serenaders, Wingy Manone's Cellar Boys, and the Frank Melrose records. Not entirely lacking in relatively simple, easily remembered head arrangements, records like these still primarily focused on a multi-chorus melodic statement and polyphonic elaboration of the tune being recorded. Black groups were more likely to use a blues progression and theme as the basis for such improvisations, while white bands usually chose a popular song of the day.

On recordings like these, often quite distinctive sounding instrumentalists combined their personal and highly expressive instrumental timbres and techniques in practiced demonstrations of polyphonic freedom.

They seemed, when compared with arranged dance band records at least, to capture jazz's freedom from the tyranny of precise melodic repetition and from the regimentation of written arrangements. But despite their celebrated (and often bitterly condemned) polyphonic freedom, such records actually explored (and re-explored) a limited musical and show business territory. The impression of creative freshness and novelty initially produced by these groups depended upon the discovery of their particular polyphonic texture. Despite their efforts to find interesting new material, different instrumentalists, and variety in tempi, such groups recorded and re-recorded jazzband polyphonies.

A substantial number of less arranged records by black outfits manipulated minstrel and vaudeville effects. Novelty and hokum records made up a large proportion of those made by Jimmie O'Bryant's Washboard Bands (with either W. E. Burton or Jasper Taylor on washboard), Jimmie Blythe's Ragamuffins (Jasper Taylor's washboard on tunes like "Messin' Around" and "Ape Man"), Jelly Roll Morton's Steamboat Four and Stomp Kings (with "Memphis" playing the comb and W. E. Burton, the kazoo), and J. C. Cobb and his Grains of Corn (also with W. E. Burton on kazoo). On a few of the Johnny Dodds sides, minstrel show monologues set the tone of the three-minute recording.

White recording executives probably encouraged such overtly primitivist instrumentation. White A & R men like Frank Melrose, Jack Kapp, and Ralph Peer often discussed recording sessions ahead of time with black band leaders, searching for marketable compromises between what the musicians seemed prepared to play and what the company believed would sell. Even when blacks were in charge of the recording session, company pressures to produce identifiably racial music still had both good and bad consequences, providing a market for black musicians but also encouraging racial stereotyping.[44]

But greater musical sophistication in jazz seems in some cases to have evolved in tension with the more overt show business solutions to the problems of generating musical excitement. The records of Jelly Roll Morton's Red Hot Peppers, nineteen of which were made by the Victor Talking Machine Company in Chicago from September 15, 1926 through June 10, 1927 (after which Morton recorded in New York), created combinations of strikingly arranged passages with polyphonic ensembles, breaks and solos.[45] On these records, more than on his earlier ones, Morton focused on instrumental effects in the jazzband tradition, mixing carefully conceived head arrangements with polyphonic improvisations

on his own original compositions. Morton's critically celebrated compositions were recorded by a handpicked core of practiced New Orleans-style
improvisors, but three of them were then working in Chicago's large
black dance bands and therefore accustomed to reading arranged passages as well as improvising. They skillfully interpreted Morton's original
compositions, which featured elaborate introductions and codas, choir-
like harmonies for the wind instruments, two-bar harmonized as well as
individual solo breaks, unison and harmonized riff figures, and a loose,
swinging jazz beat.

Morton's Red Hot Pepper recordings blended a remarkable precision
in the performance of the leader's arrangements with relaxed polyphonic
ensembles. Morton fashioned an unprecedented compositional and tex-
tural originality out of the seven-piece jazzband tradition, which had
moved from New Orleans to Chicago. Although he had used the saxophone as a lead instrument before 1926, Morton avoided the two-
saxophone section; he arranged, however, lovely harmonized passages for
clarinetists Darnell Howard and Barney Bigard on "Sidewalk Blues" and
"Dead Man Blues." He even had Omer Simeon double on bass clarinet,
an unusual instrument in jazzband music.

Jelly Roll Morton recorded both musically focused and novelty-
oriented records and was more likely than Oliver to reach for fairly
blatant vaudeville effects when recording entertainment-oriented sides:
"Hyena Stomp" spotlighted contagious belly laughter, and "Billy Goat
Stomp" immortalized rhythmic bleating. Records like these returned
music to the service of stage humor. The majority of his Red Hot Peppers
sides, on the other hand, substantially enlarged the territory of musically
focused cabaret entertainment.

The interweaving of polyphony and arranged passages also charac-
terizes some of the more rhythmically urgent and instrumentally naive
records of the young rebel white Chicagoans, who sometimes produced a
wildly disorganized sound on their recordings. Muggsy Spanier's Bucktown Five used plenty of arranged passages, notably on their recording of
"Really a Pain," an elaborately contrived arrangement whose title might
have indicated Mel Stitzel's enjoyment at writing a very complicated
piece, or the musicians' feelings about recording it, or both.[46] McKenzie
and Condon's Chicagoans integrated some basic structural signposts like
"the flare," an ensemble punctuation played at the first ending of a
chorus, "the explosion," a brief polyphonic ejaculation at the end of the
first eight bars of a thirty-two-bar popular song, and "shuffle rhythm," an

implication, often during the bridge, of double time by the rhythm section, while reveling in the wildest (nearly anarchic) sorts of polyphonic improvisation, what guitarist Marty Grosz has called "the manic frenzy of the ensemble imbroglio." These very young white musicians brought an unprecedented rhythmic urgency and carefree (nearly careless) individualism and solo abandon to their best records, but Bud Freeman, Mezz Mezzrow, and Wingy Manone recorded solos in Chicago which demonstrated a groping for improvisational control. In the 1920s, at their early stage of musical development, their rough tonal production, angular solo lines, and dissonant harmonics worked best in the ensemble mode.[47]

Frank Teschemacher contributed the more arranged touches to the Chicagoans' records: the introductions, notably the harmonized passages on "Nobody's Sweetheart," and "China Boy," and the unusual 6/4 meter on "Liza" by McKenzie and Condon's Chicagoans. "Jazz Me Blues" recorded by Frank Teschemacher's Chicagoans even features written arrangements for three saxophones, an unusual attempt among the white rebel musicians to integrate dance band concepts with improvisations. The same tune recorded by the Charles Pierce Orchestra in April 1928 (Pm 12640) included a marvelously written variation for three saxophones, played with urgent forward movement.[48]

The most widely admired ensemble approach featured one or more exceptionally powerful, controlled solo instrumentalists within a polyphonic context. Louis Armstrong's Hot Five and Hot Seven records, for example—recorded beginning on November 12, 1925—combined technically limited but emotionally powerful polyphonic ensemble players like Edward "Kid" Ory and Johnny Dodds with Armstrong's flamboyantly dramatic soloing on a series of harmonically simple, often blues-based tunes. The limitations of the other instrumentalists and, as James Lincoln Collier[49] has noted, the carefree simplicity of the underlying structures enhanced the impact of Armstrong's astounding virtuosity and inventiveness. "West End Blues," for example, combined Armstrong's technically advanced, dramatic playing with an unadorned blues otherwise performed in a relatively undistinguished fashion.

The cornet/trumpet star was not the first to take solos on records: members of white New York jazz bands had recorded solos as early as 1922, while they also had been part of NORK's pioneering Gennett sides one year later; Beiderbecke's solos on the Wolverines' "Tiger Rag" and "Royal Garden Blues" (among others) in 1924 had announced on record some impressive possibilities for jazz solos. But Armstrong set newer and more flamboyant standards. His solo work fused comedy and melodrama with brilliant musical invention. His influence is everywhere on the Chicago jazz records of the latter half of the 1920s, as other horn players tried unsuccessfully to capture his expressive eloquence. Among South Siders, Lee Collins, Jabbo Smith, and Reuben "River" Reeves—the last-named largely ignored in jazz histories—strove to emulate Armstrong's style, while Wild Bill Davison, Wingy Manone, and Max Kaminsky followed suit among the whites playing lead horn in Chicago.

Reuben ("Red" or "River") Reeves and His River Boys made a series of recording sessions in 1929 which framed the trumpeter within a polyphonic jazzband. Reeves, a skilled trumpet soloist inspired by Armstrong, had been born in Evansville, Indiana, and earned a master's degree from the American Conservatory of Music before joining Dave Peyton's Grand Theater pit orchestra in 1927. He went on to play with Peyton's Regal Theater Orchestra and taught music at Wendell Phillips High School before cutting his River Boys sessions during the summer of 1929.[50]

Reeves possessed a remarkable trumpet technique, power, and range; Dave Peyton touted him as the heir to Armstrong's throne. But his small band records suggest that the educated musical technician had technically mastered many of Armstrong's mannerisms but not the spirit of his individual style. "River Blues," "Parson Blues," and "Papa Scag Stomp" present flashy displays of trumpet, piano, and clarinet technique, but often become satirical in spirit: Reeves's version of Armstrong's "shake" becomes a horse-like whinny, while falsetto vocals communicate the comical, ever so slightly condescending attitude which educated musicians often brought to playing in even a sophistcated blues style.

Jabbo Smith, another Armstrong follower who recorded in Chicago late in the decade, came closer than Reeves to replicating Armstrong's solo range, technique, timbre, and spirit within the polyphonic jazzband approach. Smith, moreover, recorded with superb musicians—banjoist and guitarist Ikey Robinson, clarinetist Omer Simeon, tubist Hayes Alvis, and the marvelous pianist Cassino Simpson—who played well in the Hot Five style. As Jabbo Smith's Rhythm Aces, these musicians made exciting recordings of "Ace of Rhythm," "Sau Sha Stomp," and "Little Willie Blues." A reckless exhibitionism led Smith into solo stunts from which he sometimes had trouble escaping satisfactorily, and this same quality precluded his reaching the emotional depth of Armstrong's playing.[51]

Framing more than one impressive soloist within the jazzband model sometimes stretched this approach to its limits, particularly when Armstrong combined with Earl Hines, the second-most acclaimed soloist on Chicago's race records from the twenties. Hines's path-breaking talent was not immediately apparent, however. He first recorded with Johnny Dodds's Black Bottom Stompers in April 1927, one month before he played on Armstrong's Hot Seven sessions. Hines had played in leading South Side Chicago clubs since 1923. He had studied piano formally in Pittsburgh before moving to Chicago and he paralleled Armstrong in instrumental mastery. On the early records he made with the clarinetist and the trumpeter, Hines tailored his work to the older New Orleans style, playing more accompaniment than solos, his outstanding abilities hidden under the old fashioned polyphonies.

In December 1928, however, Earl Hines was finally given the occasion to record on his own. He produced eight solo numbers for the QRS company in New York and four more for Okeh in Chicago. These records reveal Hines's light, precise, and powerful touch, his sweeping keyboard control, and his lightning coordination. He produced memorable interpretations of "Blues in Thirds," "Monday Date," and "Off Time Blues."

Hines had developed, through long years of study and practice, unprecedented rhythmic freedom. He frequently and dazzlingly varied the traditional stride left-hand patterns with on- and off-the-beat walking tenths, rolling boogie woogie eighths, and irregular rhythmic accents. With his right hand, Hines played melodies and improvisations in octaves (while commenting and decorating with the middle fingers of the right hand), a pattern known as his "trumpet-style," for its resemblance to the solo voice clarity of an improvising trumpet. Most impressively, he mixed the great number of patterns of which each of his hands was capable into a brilliant variety of combinations, many of which suspended, delayed, anticipated, and interrupted the underlying rhythmic pulse.[52]

The records made in Chicago by Louis Armstrong and Earl Hines from June to December 1928, under the name of Louis Armstrong and His Hot Five, Louis Armstrong and His Savoy Ballroom Five, and Louis Armstrong and His Orchestra, stretched the conventions of jazz band performance of the time and sometimes broke with the form of polyphony plus solos. On such records as "Muggles" (an early slang term for marijuana) and "Weather Bird," the two virtuosos introduced a new

range of performance possibilities in the interaction of just two musicians. By combining two (and potentially more) exceptional solo instrumentalists, they demonstrated a new sequential approach to instrumental
interaction. On the out-chorus on "Weather Bird," Armstrong or Hines
called out or responded to solo figures which seemed to unfold in conver-
sational dialogues. The empathy and sensitivity required to complement
one another were enriched by their ability instantly to seize and creatively
extend each other's solo lines. In the process, Hines and Armstrong
created fresh elements of drama and suspense.

Clarinetist Jimmy Noone's Apex Club Orchestra, which cut sixty-
three numbers for Vocalion during the twenties, also featured two elo-
quent instrumental soloists in an interestingly arranged and loosely
played setting. Noone's records were influential in enhancing the role of
the clarinet as a lead and solo instrument. The group's instrumental
sound focused attention on Noone's beautifully clear, centered, and
round-toned woodwind style, juxtaposing his warmer, lighter sound to
the brassy New Orleans tradition and featuring an interplay between
Noone, a baroque improviser, and Joe Poston, a non-improvising alto
saxophone section man who carried the melodies. With the technical
refinement of Earl Hines (later replaced by Alex Hill and Zinky Cohn) at
piano and a slick, fast rhythm section led by drummer Johnny Wells, the
longest-lasting member of the group, Noone's band epitomized sophisticated cabaret jazz.[53]

Noone's solos and duet embroideries featured a variety of thought-
fully manipulated instrumental techniques. Typically, as in his recording
of *San,* Noone's solo motifs began in the clarion register and descended
deep into the chalumeau before working their way upward once more.
He often featured staccato-tongued broken chord figures which danced
over the changing harmonic progressions. Noone liked contrast, and for
nearly every one of his sharply tongued phrases, a very lightly tongued or
slurred figure soon followed. Similarly, he easily jumped large intervals,
as his signature "Apex Blues" and his up tempo "El Rado Scuffle" and
"Chicago Rhythm" indicate. During the slower numbers, after establish-
ing a long bathetic half or whole note by hitting it squarely in the center
with his wonderfully focused tone, he would bend it and work it over
with a widening vibrato, a parallel on the clarinet to Armstrong's shake,
so wide a vibrato as to burst into one of his dissonant, birdlike trills that
ranged from a minor second to a minor third. Various ingredients of
Noone's style were absorbed by Albert Nicholas, Barney Bigard, Omer

Simeon, Cecil Scott, Jimmy Dorsey, Benny Goodman, Buster Bailey, Frank Teschemacher, and Mezz Mezzrow.

Noone's Apex Club Orchestra created its own relaxed, informal but elegant after-hours fusion of music and show business. Band members or featured vocalists sang earthy double-entendre lyrics on numbers like "It's Tight Like That," "Let's Sow a Wild Oat," and "My Daddy Rocks Me (With One Steady Roll)" and eerily juxtaposed musical high spirits with songs about suicide ("Ready for the River" and "Four or Five Times"). Two numbers, "Wake Up Chillun, Wake Up," and "On Revival Day," intermixed the sounds of the spirituals with red hot twenties stomps like "Oh, Sister! Ain't That Hot?"

The records made in Chicago by Bix Beiderbecke and the Wolverines also combined solo virtuosity with polyphonic ensembles. The relative instrumental naiveté of the other Wolverines enhanced Beiderbecke's sound in much the same way the Hot Fives and Sevens had enhanced Armstrong's. But Beiderbecke's approach to improvisational solo freedom was far less flamboyant than Armstrong's. Beiderbecke displayed his exceptional abilities within the ensemble, during breaks, and in his famous chorus-long solos. The beautiful bell-like clarity of his tone and his flawless musical discretion, lifted his undistinguished studio colleagues into history. He attempted less than Armstrong, not just because of his retiring personality. Few of the white Chicago jazzmen, and certainly none of those who played on college campuses, had ever been required to play the sort of solo features pioneered by Armstrong in cabaret floor shows. Wild Bill Davison's extensive solo features on the vaudeville stage with the Benny Meroff Orchestra were exceptional among the white jazzmen. They usually played for dances and for each other, adopting less of the dramatic flamboyance of the South Side soloists and all but eliminating vocals, comedy routines, religious references, and the like.[54]

Armstrong's seven-man group which recorded in 1928 as the Hot Five and the Savoy Ballroom Five used some effectively arranged passages to complement and highlight their matchless solos. The outstanding examples were "St. James Infirmary Blues," "Beau Koo Jack," and "No One Else But You." The most complex was Alex Hill's "Beau Koo Jack," an up-tempo jump or stomp number with a three-theme structure complicated by an intricate introduction, two modulations, breaks, and a breath-takingly challenging, thirty-six-measure concluding ensemble arrangement. It was an Armstrong *tour de force;* the trumpet star played a

thirty-two-measure solo and immediately jumped into thirty-six measures of taxingly arranged ensemble lead. The arrangement featured long
ensemble lines which sounded like (and might once have been) improvisations, scored for the four frontline wind instruments. Such scored
passages may well account for Armstrong's selection of Jimmy Strong on
clarinet and Fred Robinson on trombone in place of Johnny Dodds and
Kid Ory. Whatever their demonstrated shortcomings as soloists (neither
possessed the rugged, untutored power of the musicians they replaced)
Armstrong's new sidemen could read and therefore more swiftly master
the written sections.[55]

"St. James Infirmary," a stagy musical drama in minor mode, elicited
little solo inspiration from either Armstrong or Hines but did feature
saxophonist/arranger Don Redman's softly haunting, arranged ensemble
line in the background behind Fred Robinson's trombone chorus. Redman then brought his long, loping, written ensemble line to the foreground for a rousing last chorus which fittingly contrasted a new technical sophistication with an old-fashioned tune.

Another group of competent, versatile black Chicago musicians, who
were often good if not exceptional soloists, focused on playing well-
conceived small band arrangements. Records by the rarely mentioned
Tiny Parham and His Musicians, who recorded extensively for Victor
from 1928 through 1930, presented the most fully arranged small group
sound of the era and yet succeeded by not becoming too ambitious or
overly elaborate. Hartzell Strathdene Parham, a tall, heavy man who had
been born in Winnipeg, Manitoba, in 1900 and raised in Kansas City,
came to Chicago in 1925 after touring on the Pantages circuit with his big
band. After co-leading a band with violinist Leroy Pickett, Parham then
led his own group at La Rue's Dreamland, the Sunset Cafe, the Granada,
Merry Gardens, El Rado, and the Golden Lily. Late in the decade, all of
these clubs were heavily patronized by whites. Al Quodbach's Granada
on Cottage Grove Avenue, for example, had featured Paul Whiteman,
Guy Lombardo, and Ted Weems before Parham took over in 1931.

Parham also worked as an arranger under Mayo Williams of Paramount Records, who remarked that he was "a hell of a piano player, but
he couldn't get 'down to earth' on blues [as a performer]." Parham's
particular talent, however, was described by a Paramount salesman:
"You could sing it (a blues) one time; he could play it back; he'd write it
the next time."[56] Parham favored a relatively refined, arranged jazz
sound, recording with a seven-to-eight-piece band that featured New

Orleanian trumpeter Punch Miller as get off man. He built intelligent, challenging, and not overly fussy arrangements on original numbers like "The Head Hunter's Dream (An African Fantasy)," "Voodoo," "Jungle Crawl," "Lucky '3–6–9,'" "Nervous Tension," and "Bombay."[57]

"Voodoo," recorded on February 1, 1929, centers on the possibilities of a minor mode and minimal harmonic movement. After a dramatic rubato piano introduction in Db major by Parham, the violin and saxophone state a simple unison riff figure in Db minor over an eight-bar suspended, unchanging tonic chord, sharply delineated by an ostinato bass figure based on the first three intervals of the scale and played in unison by Quinn Wilson on tuba and Parham's left hand. A brighter ray of harmonic movement provides an abbreviated "bridge" of four bars, before a return to the ostinato pattern and modal suspension that round out the sixteen-bar frame. The solo instruments concentrate on full tonal production rather than complex lines, encouraging the listener to enjoy the underlying mystery inherent in that eerie minor figure in the bass. (Parham worked extensively in vaudeville and movie theaters as well as the large cabarets, and many of his records show a penchant for theatrically dramatic moods—mysterious, exotic "Bombay," elegantly decorous "Cathedral Blues," and the slickly prancing "Washboard Wiggles"). His "Dixieland Doings" adopts an analytic stance toward jazz band polyphony that still preserves some of the genre's excitement. Parham's carefully crafted arrangements often made up for his group's lack of a creative soloist of Armstrong's stature.

As the decade waned, a more arranged sound competed with the earlier improvised polyphonies. The number, size, and popularity of cabarets had encouraged the recording of small jazzbands; but, as the courts moved to close most of Chicago's most prominent cabarets in 1928 and 1929, a process considered in more detail in the following chapter, those jazzmen who remained in the city began to compete for jobs in the dance halls and the few remaining, large cabarets. This process encouraged the development and recording of bigger, arrangement-reading jazz bands.

The large black dance and theater orchestras of Charles Elgar, Erskine Tate, Charles Cook, Sammy Stewart, Carroll Dickerson, and Earl Hines produced complexly arranged big band music with and without creative solos. For example, Erskine Tate's twelve-piece Vendome Theater Orchestra, which played light concert hall classics as well as accompaniments for vaudeville and movies, recorded a very hot, hard, charging romp entitled "Static Strut" with Louis Armstrong on May 28, 1926.[58] While the record features two solo choruses by Armstrong on an harmonically straightforward sixteen-bar pattern, this number shows attention to arrangement. The two trumpets play a riffing stop-time figure in the introduction and verse, while the three saxophones chord on whole notes. Tate then shifts the lead to his saxophone section over prominent woodblock counter-rhythms by percussionist Jimmy Bertrand. The woodblocks then recede, and the muted trombone leads in dialogue with the saxophone section. A four-bar arranged modulation from C major to E♭ major introduces Armstrong's ecstatic solo with rich stomping piano accompaniment by Teddy Weatherford, who then makes an unprepared modulation back to C in order to solo for sixteen bars without rhythm section accompaniment. Weatherford is then followed for a chorus by an unknown alto saxophonist, and Stump Evans concludes the solo choruses with a slap-tongued statement on his baritone sax. Armstrong then returns to rip off a stop-time figure in the first four bars of the last chorus, answered by a hot orchestral polyphony in measures five through eight; the same configuration repeats before the performance struts into a four-bar coda, again divided into rhythmically contrasting two-bar effects.

Throughout "Static Strut," the musicians are kept on their toes by changing patterns of rhythmic breaks, particularly during the last eight bars of Armstrong's second solo chorus, when four differently arranged two-bar breaks follow one upon the other. Exciting Chicago jazz records like this one represented just one form of cross-fertilization between the large theater and dance bands and the cabaret soloists. In this context, virtuoso instrumentalists like Armstrong and Weatherford were called "get-off men" for their ability to improvise. If the get-off man couldn't read music, another musician playing the same instrument would be hired to read the arrangements.

The power of this style was even more clearly delineated by Carroll Dickerson's orchestra. Dickerson's crew also recorded with Louis Armstrong and Earl Hines as featured soloists on "Symphonic Raps" and "Savoyagers' Stomp." Dickerson had played long runs as band leader at the Sunset Cafe, toured in vaudeville with his big band, and played extensively in the Savoy Ballroom starting in 1928. He was just one of several South Side musicians whose careers spanned the improvisatory and "legitimate" musical worlds. Dickerson's recording of "Missouri Squabble" did without great soloists but still managed to create exciting jazz in an interestingly arranged rhythm number in popular song form, recorded for Brunswick in May 1928.[59] The well-rehearsed arrangement, which opens with an eight-bar chromatic progression, carries the performance nicely, and its extended harmonized lines for the saxophone section in both the A and the B sections demonstrate the power and excitement generated by replacing the soloist with a well-rehearsed choir of saxes attacking a written variation on the melody and chords with a finely honed rhythmnic identity. A well-played chorus-long, linear variation carried even more punch than a lone solo instrumentalist might have.

The relatively few big band sides from twenties Chicago underscore the close relations between jazz bands and larger orchestras. Charles Cook's Dreamland Orchestra included a core of jazz-oriented musicians who heated up the orchestral arrangements and sometimes stepped out together to play as a separate small jazz band. Cook reedmen Jimmy Noone, Joe Poston, Clifford King, and Jerome Pasquall shared jazz sensibilities with cornetist Freddie Keppard and banjoist Johnny St. Cyr. These men brought a wild stomping spirit to the Dreamland Orchestra's recordings of "Hot Tamale Man" and "High Fever" on Columbia. Keppard, Noone, Poston, St. Cyr, trombonist Fred Garland, and pianist Kenneth Anderson also emerged from the dreamland Orchestra to form Cookie's Gingersnaps, a marvelous hot jazzband which both improvised and featured arranged effects on its own recorded versions of "High Fever," "Here Comes the Hot Tamale Man," and "Messin' Around."[60]

The Earl Hines Orchestra, formed as a rehearsal band while the star pianist was appearing with Jimmie Noone's Apex Club Orchestra, recorded in 1929. This group sought a synthesis of the arranged, twelve-piece orchestral sound with the hot, agitated Chicago beat. As Gunther Schuller has noted, the Hines orchestra failed to achieve an identifiable sound because it used too many different arrangers, and the band's records sometimes sound under-rehearsed.[61] But these records still mark an important step in the decade's steady movement toward show business and musical sophistication. The fact that so many sidemen not only wanted to but were capable of writing arrangements indicates what had been happening to jazz music in Chicago.

Hines went after the best of Chicago's new generation of jazz musicians, young men like reedman Cecil Irwin and brass bassist Hayes Alvis. These ambitious jazzmen sight read skillfully, soloed smoothly, and had studied arranging. Of the records by Hines's 1929 aggregation, "Chicago Rhythm," primarily a rhythmic vehicle with little in the way of remarkable harmonic or melodic content, was the most exciting.[62] It features an arrangement of a stock thirty-two-bar popular song in which a swiftly moving train-whistle introduction leads into a full chorus, harmonized, saxophone choir statement of the melody. The trumpet choir shares the second chorus, with a sax solo on the bridge, before an unusually arranged eighteen-measure interlude. Hines then plays a romping, swooping solo on the A section of the ABA song form before the entire band plays it out, concluding with a skyrocketing fanfare.

Ben Pollack and His Orchestra recorded elaborately arranged jazz performances while still retaining informality and excitement. Pollack, a Chicago drummer who had played and recorded with the New Orleans Rhythm Kings, subsequently worked in California where he assembled an exceptionally talented band which included Benny Goodman, saxophonists Gil Rodin and Fud Livingston, and trombonist Glenn Miller. The group played in Chicago in the Venetian Room of the Southmoor Hotel and the Blackhawk Tavern beginning in 1926 and recorded for Victor. On "He's the Last Word," a popular song ("He can't talk, but on a moonlight walk, he's the last word") sung by Hannah (Mrs. Jack Dempsey) and Dorothy Williams, Glenn Miller's demanding, sophisticated arrangement wedded the white tradition of dance band arranging to hot, improvised jazz. Using impressionistic whole tone scales, dense, odd chordal progressions, parallel and chromatic motion, and unusual modulations (D maj to E min to B_b minor to E maj to D_b min and back to E min) Miller left no doubt of his voice-leading skills and theoretical sophistication. Benny Goodman is the featured soloist and improvises confidently, combining some startling Teschemacher-inspired pyrotechnical leaps into the upper register with a precocious instrumental and improvisational control.[63]

Chicago's jazz records from the 1920s, therefore, offer a variety of approaches to the combining of the oral/aural musical traditions that were brought to the city with a lively and competitive market for musical entertainment. These strategies were intended to catch the attention of cabaret audiences and to create identifiable musical reputations that would sell phonograph records. In the process, the leading musicians working in twenties Chicago created a musical genre whose appeal has endured well beyond the special jazz age to which it made such a major contribution. As the foundations of Chicago's jazz age crumbled in the late twenties, jazz music swiftly adapted to new economic, political, and cultural conditions, leading the way toward Swing.

6

"Syncopated Threnody": The End of Chicago's Jazz Age

THE END OF Chicago's 1920s Jazz Age could not, by any means, put an end to jazz in Chicago. During the decade, the musical and entertainment power of jazz had been more compellingly demonstrated than ever before. Musicians like Art Hodes, Punch Miller, Bud Jacobson, Milt Hinton, and Franz Jackson would follow the musical paths traced by the Jazz Age greats and would explore their own jazz strategies, but they would enjoy few of the supporting economic, political, and social conditions which had made jazz the central cultural experience of the 1920s.

The quantity, intensity, and type of jazz in Chicago did change in the late 1920s as the city's leading jazz musicians moved on—mainly to New York City—to participate in new forms of popular music. Eventually, economic, musical, and political conditions would converge again, creating a dominant new design called the Swing Era to rival the Jazz Age; but in that transition much was lost. Chicago's leading jazz cabarets had provided a powerfully concentrated, highly charged jazz experience. The leading black-and-tan cabarets had presented the best black jazz musicians, singers, and dancers of a talented and ambitious generation to inter-racial audiences that were packed close to the bands and to each other on small dance floors and around tiny tables. By comparison, the emerging Swing Era would inherit the racial segregation of big city dance halls of the twenties, which had allowed only all-white audiences, would confine jazz soloists to arrangement-reading dance bands, and would dilute the musical experience in large open spaces filled with self-absorbed social dancers. Many who lived through both periods never forgot the intense excitement of Chicago's jazz cabarets during prohibition.

A number of factors contributed to the decline of Chicago's historic Jazz Age. The buoyant, heady, and giddy sensibilities of twenties Chicago had been intensified by an unprecedented historical combination of a booming economy, machine politics, rebellion against prohibition moralism, and increased inter-racial awareness. The stock market crash of October 1929 removed one of the most important outlets for these sensibilities by cutting out most of the discretionary, leisure time income of many urban workers and bureaucrats. Many of the cabaret entrepreneurs were forced out of business. Of "black Tuesday" on October 29, 1929, when the market crashed for the second time and failed to recover, *Variety* headlined "Wall Street Lays an Egg" and a week later declared that "the bottom [has] dropped out of hilarity."[1]

But Chicago's urban reform moralists and racial segregationists had long since served notice against the continued existence of the black-and-tan cafes. South Side cabarets suffered a slow death during the late twenties. The historical circumstances which had attracted the best musicians to the city began to change. Many of the most talented jazz artists—those who had created musical excitement in what were sometimes dismal basement speakeasies—left Chicago in 1927 and 1928.

Observers close to the scene reported that adverse conditions had begun to undermine Chicago jazz as early as October 1925. That year, Jack Lait, a pioneering Chicago night-life reporter, in a comment on the revived power of urban reformers under Mayor William Dever, had declared that the city's fabled cabaret era was dead. He said that the town was now dominated by its "paunchy merchants and axe-faced reformers." In its normal seasonal cycle, the cabaret business wilted in the summer but revived in the fall. But the autumn of 1925 found South Side black-and-tans abolishing their cover charges in order to "coax attendance." Just over a year later, *Variety* noted that the entire night-club business in Chicago was "a slowly sinking trade."[2]

The fate of Isadore Shorr and Joe Gorman's Entertainers Cafe on the South Side stood as a reminder of what the wrong combination of urban reformers, Democratic politicians, hostile court justices, and racial segregationists could do to the entertainment world. Repeatedly singled out for liquor violations from 1920 to 1924, the Entertainers Cafe was closed down in mid-1924 for a year and a day by the United States Circuit Court of Appeals. Part owner Joe Gorman, a white person, was jailed. He died upon leaving prison, and his widow's attempt to reopen the club was unsuccessful. The Chicago *Defender* concluded that Gorman, who was "very popular among our people" and "absolutely unprejudiced," had become a martyr to racial injustice.[3]

This continuing campaign by urban reformers against the black-and-tan cabarets spelled the beginning of the end. *Variety* had warned in late 1925 and early 1926 that the black-and-tan club scene, and the Plantation Cafe in particular, had "turned rough—too rough even for those seeking high-power ... thrills." Lait had claimed that a tough "colored element" around these "hybrid roisteries" no longer distinguished "white sight-seers and kick-seekers" from the show business people who, he claimed, had "made this cafe." In an uncharacteristic vein, Lait, usually even-handed in covering race news, reported that white people and especially white women were no longer protected. He claimed that actors and entertainers going to the Plantation Cafe in April 1926 exposed themselves "to unpleasant publicity and trouble by being in attendance."[4]

Over the 1926–27 Christmas and New Year's holidays—at the end of William Dever's tenure as mayor and just before Big Bill Thompson regained the office—Chicago police raided many South Side clubs and even arrested and fined several owner-managers, including the usually untouchable Joe Glaser of the Sunset Cafe. According to the trade papers, he and Virgil Williams, then of the Dreamland Cafe, were separately charged with "mistreating" under-age girls and contributing to their "delinquency." The charge against Glaser was pressed for one of the girls' fathers by the Board of Education. No further press coverage detailed the outcome of either case, which would have come to trial under the Thompson regime.[5]

Jazz musicians continued to do their best to attract crowds into struggling clubs, generating some of their most frenetic displays of musical heat. In April 1926, Dave Peyton reported that Carroll Dickerson's Sunset Cafe band with Louis Armstrong and Zutty Singleton was "red hot"; and he added that, with Joe Oliver's Dixie Syncopators at the

Plantation Cafe, 35th Street was likely to burst into flames. But in March 1927, with city police raids in progress throughout the South Side, the Plantation Cafe suddenly closed, putting Oliver's band out of work. The veteran band leader responded by mounting his eleven-piece band on a flatbed truck to play on street corners throughout the city, much as he had earlier on wagons when starting out in New Orleans. As the Plantation Cafe reopened and closed again with the shifting tides of city politics, the Peyton-Oliver competition for work at the club resumed. Peyton won, and Oliver, starting the jazz exodus from Chicago, left in late April on a tour planned to reach its climax at New York's Savoy Ballroom.[6]

Although local political pressures on cabaret speakeasies had intensified under Mayor Dever, the jazz clubs had still managed to remain in business. Owners had appealed municipal closing orders, often succeeding in securing writs of *mandamus* or *supersedeas* by which appellate courts ordered that the cabarets in question be allowed to remain open until final legal appeals were heard, perhaps as high as the Supreme Court. From 1923 to 1926, therefore, all cabarets had functioned in a legal limbo, pressured by urban reformers and reform-minded Chicago politicians but still managing to remain open. They were kept in business by the unwillingness or inability of the city police to discover their infractions, the preoccupation of federal agents with the manufacture and distribution rather than the retailing of alcohol, and temporary court injunctions issued either by federal appellate courts or by municipal judges that were allied with the Thompson Republicans. Even when their appeals had been exhausted the maximum punishment that they received for proven liquor violations was a one-year suspension of business.[7]

Nevertheless, legal ambiguities and political pressures did lead the most prominent black jazz businessman—Bill Bottoms—to disassociate himself from the South Side black-and-tans and move his business outside of the city limits. The revered neighborhood cabaret owner sold his Dreamland Cafe and moved his operations to Robbins, Illinois, where he opened a roadhouse that featured Darnell Howard's Orchestra, satirically named the Farm House Country Club. Bottoms did return to the South Side with the repeal of prohibition. Virgil Williams, who had run the famed Royal Gardens in its heyday and the Dreamland Cafe for a time, after Bottoms quit, also got out of the South Side cabaret business.

During the late twenties, white gangsters, who had always been a force in the Black Belt, tightened their grip on the Stroll. The Al Capone syndicate reputedly bought the Plantation Cafe. Joe Glaser had also bought into the Plantation. Ed Fox owned the Grand Terrace. As the noose of federal prohibition tightened, business fell off, and competition became violent. Bombs ripped through the Plantation and the recently renovated Cafe de Paris. (The Sunset stood untouched.) The most famous South Side cabarets became increasingly notorious as gangland properties, cutting themselves off from legitimate businessmen in both the black and white communities.[8]

Even the return of Big Bill Thompson to the Mayor's office in 1927 failed to reverse the tide. Thompson's Chief of Police publicly pledged "respect for personal liberty and no more mass arrests." But a national campaign, first carried out by federal agents against cabarets and speakeasies in Chicago, succeeded in closing all of the mainstays of the jazz age. Clubs had successfully defied prohibition as long as prosecutors remained unable to prove that illegal alcohol, discovered and then confiscated on the premises, had been sold by or even in the club. But in December 1926, not long before Thompson regained office, this legal loophole was closed when federal judge Adam C. Cliffe heard the appeal of Chicago's Moulin Rouge, Friars Inn, and Al Tierney's Town Club, three of the Loop's leading night clubs, and ruled that the Volstead Act outlawed not just public places which actually sold illegal alcoholic beverages but also "places where people carrying liquor congregate." Such clubs were now guilty of aiding and abetting customers in the public consumption of alcohol if they provided set-ups of glasses, ice, water, and ginger ale to anyone who carried liquor onto the premises in a "hip flask."[9]

Led by Mike Fritzel of Friars Inn, the three clubs appealed to the United State Supreme Court. The court was asked to refuse testimony from agents who claimed to have witnessed patrons "drinking from a bottle something that looked like whiskey." In October 1927, the Supreme Court refused to hear the cabaret owners' appeal of what *Variety,* always a supporter of the speakeasies, called the "hip" rulings. The city entrepreneurs regarded the Court's refusal to intervene in the federal crackdown as "the death knell for night club business in Chicago and perhaps the entire country."

One month later, twelve clubs were padlocked under the "hip flask" ruling. On February 7, 1928, at 1:30 a.m. on a Monday morning, federal agents conducted a blanket raid on the largest and most prominent cafes both in and out of the Loop: included were the Blackhawk, Rainbo Gardens, Plantation Cafe, Club Bagdad, Silver Slipper, Jeffery Tavern,

Rendez-vous, and Midnight Frolics. The Moulin Rouge, Friars Inn, and Town Club, of course, had already been closed. Thousands of customers were in the clubs at the time of the raids; police took the names of those who had liquor at their tables. Agents padlocked an estimated three million dollars in property. *Variety* began to talk about Chicago's jazz cabarets in the past tense:

> Chicago was once the hottest cafe town in the United States, famous for sizzling music, torrid night life, a great little spot for the great little guys. But that's history now. Night by night it gets tougher for the cabarets.

The era of the flashy cabaret had thus been brought to a close (but bequeathed another likely source for the jazz term "hip" or "hipster," as a fanatic of jazz, alcohol, and cabarets who defiantly carried a flask hidden in a hip pocket).[10]

These raids did not come as a complete surprise to cabaret owners. According to *Variety,* about a month before the appeal of a specific cabaret was to be heard, "the cafe affected usually cuts down on its operation costs and entertainment." One week before the padlocks were applied, "the places clean up, dispose of all equipment and voluntarily close up shop." Entrepreneurs also discovered that the perpetual threat of federal padlocking caused insurance companies to cancel their policies since a padlocked, empty club became a risk for unexplained fires which might bring insurance reimbursement.[11]

Those few cabaret entrepreneurs who managed to remain in business were too broke even to open on the night of the Dempsey-Tunney fight. Most of them decided that alcohol was too risky and turned to gambling. The South Side's Granada Cafe, at 65th and Cottage Grove, reserved for a white clientele and owned by Al Quodbach, pioneered a new compliance with the Volstead Act by prohibiting liquor in and even around the premises, hiring doormen to check customers for hip flasks. At the same time, Quodbach aimed to capture a large number of customers, each of whom would spend less than the butter-and-egg men of earlier times but still enough to make a profit. If money was no longer to be made by milking the wealthy with overpriced prohibition alcohol and exorbitant cover charges, then club owners could turn to a higher volume of moderate spenders. But in the process, Chicago's jazz cabarets began to transform themselves into large streamlined clubs which were sometimes difficult to distinguish from dance halls. By early 1929, Dave Peyton reported that:

> Most of the night clubs using small orchestras have been closed in Chicago on account of prohibition violations, making it pretty tough on our musicians, who were in demand in the places patronized by the elite with money.[12]

The groundswell of sensibility about urban leisure time, which had been reported in the press as early as 1904, was far too powerful, however, to be completely arrested by the depression and federal prohibition. Though most of the prominent jazz roisteries of the twenties had disappeared, the cabaret business survived on a smaller and far less remunerative scale. Federal prohibition never touched an enormous number of more modest "speaks." According to *Variety,* "16,000 Beer Flats in Chicago" continued to vie for the cabaret customers. Living room and dining room cabarets lined South State Street between 43rd and 55th streets well into the Depression. Chicago under Big Bill Thompson became once more a wide open town, even if deprived of its most famous Jazz Age cabarets. Chicagoans could still drink in secluded spots throughout "the great Chicago neighborhood territory," but thanks to federal prohibition agents and the depression, not to flashy shows which included jazz bands, dancing girls, and comedians.[13]

The surviving "low rent spots" reverted to the old cafe/saloon entertainment policy—a piano player (Art Hodes played in the Rainbow Cafe, an upstairs apartment on West Madison Street) and increasingly to mechanically reproduced music. On the South Side the small neighborhood speakeasies encouraged the further development of the style known as "boogie woogie" that was performed by pianists like Jimmy Yancey, Albert Ammons, Clarence "Pine Top" Smith, Meade Lux Lewis, Clarence Lofton, Cow Cow Davenport, Montana Taylor, Charlie Spand, Wesley Wallace, and Will Ezell. Boogie woogie featured a pulsing left-hand ostinato which kept the customers dancing.

Boogie woogie had developed in the southern logging camps and migrated to Chicago along with southern workers, and the smaller neighborhood speakeasies encouraged its development. Boogie woogie's relative lack of technical sophistication reflected the retreat from the high pressure world of Chicago's leading cabarets of the twenties. The boogie-woogie pianists who recorded in the 1920s only made a few records late in the decade, just before the crash, and lived and worked thereafter beyond the awareness of most white fans and the jazz promoters who had relied heavily on the black-and-tan cabarets to locate black musical talent.[14]

Actually the crash and depression did not destroy all of the white-owned black-and-tan cabarets, although they were "scarce as hen's teeth" by June 1932, according to the *Defender*. In late December 1928, Ed Fox, one of the original co-owners of the Sunset Cafe, had opened the Grand Terrace Cafe on South Parkway and 40th Street. With a stunning floor show of light-skinned dancers directed by the experienced Percy Venable and a ten-piece orchestra led by pianist Earl Hines, the Grand Terrace survived the depression, thanks in part to gangland backing and to one of the first direct broadcast radio wires from a night club. Ninety-five percent of the customers were white.[15]

The Sunset Cafe also survived the crash and continued to do successful business at 35th and Calumet with a clientele described by the *Defender* as "98 per cent white, mostly out-of-towners." The club was "still hitting" with peppy song and dance numbers from black performers; the newspaper praised these acts but said little or nothing about the bands which appeared there.

During the years between the crash and repeal, musical praise from the black press was reserved for a select few musicians who had already proven themselves during the twenties. Clarinetist Jimmie Noone, for example, seems to have survived better than most of the other leading jazz musicians from the South Side. Playing what the *Defender* called "soft, scintillating, sweet" jazz, soft but hot, and "pouring oil on the boys of melody when they visit," Noone worked steadily at the Apex Club until it closed in February 1929. Thereafter, he resurfaced, still with his sextet, at the Ambassadore Club, at the Barone Night Club, and, most prominently at the El Rado, an important jazz club that opened in July 1929 at 55th and Prairie Avenue. The El Rado attracted a predominantly black clientele, which gave Noone a rousing farewell party in May 1931 when the clarinetist moved temporarily to New York's Nest Club. When Noone returned to Chicago in September, he played extensively in the Loop and on the North Side before returning to the Club Dixie, the Lido, and the Midnight Club.[16]

Clarinetist Johnny Dodds also defied the odds, but in a different manner, continuing a long series of engagements through prohibition at Bert Kelly's Stables, the Three Deuces, the New Plantation, Lamb's Cafe, the 29 Club, and the New Stables, in all of which the audiences were overwhelmingly white. He usually headed a traditional jazzband that included Charlie Alexander, Natty Dominique, and his brother Warren. Together, these musicians played a very traditional New Orleans jazz band music with great success, spreading the polyphonic influence into the thirties among white musicians like Yank Lawson, Billy Butterfield, and Bob Haggart of the Bob Crosby band. Dodds was able to buy an apartment building.[17]

But despite such instances of remarkable endurance, the federal crackdown on the public consumption of alcohol devastated the Chicago cabaret scene: 250 cabaret entertainers and 200 musicians had lost their jobs by May 1928, thanks to the "hip flask" ruling. Many, like Benny Goodman, Louis Armstrong, and Eddie Condon, found themselves at a stage in their careers where working for a few dollars and drinks in an obscure neighborhood apartment was unacceptable.

The cabarets had furnished the leading stages for jazz, but there had been others. Some of the more technically advanced Chicago jazzmen of both races might have turned from small cabaret bands to the movie theater pit orchestras, which had long offered performance opportunities, job stability, and technical training to many jazz musicians. But the movie theater pit orchestras were also eliminated during the last third of the decade, depriving Chicago's young jazz musicians of important career possibilities, and the Jazz Age of another major foundation. Once more, the problem was national rather than local in scope. Chicago theater musicians saw serious problems approaching from New York at least as early as 1925. The arrival in Chicago during the summer of 1927 of new technologies, marketed by Vitaphone and Movietone, brought mechanically produced sound to the movies.[18]

The new sound technologies idled musicians of both races, hitting Chicago's white musicians first. Black musicians enjoyed a temporary respite but watched apprehensively as the sound systems led to the firing of orchestras in the leading New York City movie theaters. By October, a reported 1500 "straight" musicians had been thrown out of work in Chicago's theaters. For a time, musicians hoped that the capital expense of wiring movie theaters for sound would keep the new devices out of the South Side's more humble theaters. By 1928, however, even the smallest South Side theaters began to install relatively inexpensive sound systems like Photophone, Electraphone, and Orchestraphone, which used phonograph records synchronized to the films and required only one "cue boy" to operate them.[19]

After bitter conflict with both the white and Negro musicians' unions, Chicago's theater owners fired their pit orchestras; Erskine Tate of the Vendome Theater, Clarence Black of the Metropolitan, Clarence M. Jones of the Owl Theater, Lovie Austin of the Monogram, and Dave Peyton of the Grand. An avenue for a fuller integration of black musicians into the city's music business had been closed. For the time being, many theater owners replaced their orchestras with what were often elaborate electric pipe organs, another sign of the new technological forces sweeping the music business.[20]

The rise of radio and more particularly the organization of national radio networks further undermined live musical entertainment and the jazz community in Chicago. New York controlled the new world of radio music as Gotham-based radio networks like NBC and CBS were organized in the twenties. They swallowed up many local stations and forced muscians to deal with a new set of potential employers less tied to the local scene than Chicago's WMBB, owned by Andrew Karzas at his Trianon Ballroom, and WHT, which had been installed in the Wrigley Building by Big Bill Thompson, William Wrigley, and U. J. "Sport" Herman.[21]

At first, as *Variety* noted, network radio broadcasting acted as a punishing "courtroom for jazz" and encouraged "melody stuff over hot breaks and tricks." National broadcasting appealed to the mass audience by promoting a wide variety of the latest popular songs in their most identifiable melodic form. For example, in 1929 NBC presented "The Made-to-Order Orchestra," a versatile, professional group of the fastest sight-readers who could emphasize "some distinctive quality which the public can recognize, even if it doesn't understand it."[22] Since black musicians had become identified so completely with jazz, and since the white musicians' union controlled negotiations with radio studios, African-Americans did not get their share of radio jobs. Walter Barnes, the leader of the Royal Creolians and Peyton's replacement as music columnist for the Chicago *Defender* in 1929, commented bitterly:

Our folks buy, in proportion, more radios and allied equipment and victrolas than any other group and even this seems to be overlooked when it comes to giving out contracts for broadcasting. It seems to be the belief among whites that the Race is still in the cotton fields and cannot sing or play anything else but cotton songs and blues. This is a great mistake. We are music lovers and enjoy all types and forms of music.

Gradually, however, a few black Chicago bands obtained ten- or fifteen-minute spots on local stations. Barnes and his Creolians were the first blacks in Chicago to broadcast nightly, and Earl Hines became the first African-American band leader from Chicago to enjoy network coverage.[23]

The influence of network radio on Chicago jazz proved to be just one of many forces that transformed the racial and cultural outlines of the city's jazz scene. Chicago groups had come by the late twenties to depend increasingly on new national booking agencies like the Music Corporation of America, which had moved to headquarters in New York in June 1926. The new booking agencies hired few of the South Side bands. MCA did send King Oliver's Dixie Syncopators on a long road trip after the Plantation Club closed. Black Chicago bands had, of course, traveled to New York from time to time, but such transfers had been arranged between individual club owners on a relatively informal, temporary basis. Close observers of the South Side music business, like Dave Peyton, had long complained of the prejudice against blacks in the band-booking business. The rise of national booking offices located in New York only enlarged the scope of the problem.[24]

The nationalization of jazz- and dance-band booking had the further effect of solidifying the power of whites over the marketing of African-American popular music across the nation. Verona Biggs, ex-president of Chicago's Local No. 208 of the musicians' union, moved in late 1928 to organize a local black-run agency to market "small orchestras, stage talent and other amusement features" within the city. Columnist Dave Peyton urged the South Side Musical Bunch to establish separate booking agencies in every large city across the country in order to counter the advantage which white booking agencies gave to white jazz musicians. For the most part, however, these efforts came much too late—MCA had cornered the field.[25]

With most of Chicago's leading cabarets closed, the theater orchestras silenced, and new technological and organizational forces unleashed, the outlines of the Swing Era began to appear, particularly in African-American music circles, at least five years before the initial popular success of the Benny Goodman band that heralded the big band era in 1935. In 1928, Dave Peyton suggested in his Chicago *Defender* column that movie theater orchestra leaders could create some new form of entertainment with which to save themselves. The large "stage band" might replace the theater pit band. He promoted the idea, which had

been pioneered in the Loop by Paul Ash, of a large versatile stage orchestra which could both accompany the full range of vaudeville-style stage
acts and play jazz and other features as well. Such bands would now be led
by a "personality" with star quality who could entertain, banter with the
acts and musicians, and, of course, lead the orchestra. Peyton emphasized
that versatile Chicago musicians like Reuben Reeves would be able to find
employment in these new stage bands. Following the Paul Ash model,
Fess Williams and his Joy Boys were installed in early 1928 at the Regal
Theater. However, due to the Depression, the movie theater stage band,
an important forerunner of the swing bands, never became a permanent
feature in the nation's theaters, and Peyton's hopes were dashed.[26]

A few of the leading Chicago orchestras did manage to find work in
the more elaborate roadhouses of Cook County, as much of Chicago's
night life ventured into the country, where federal enforcement of prohibition was weak. The Coon-Sanders Nighthawks, locked out of the
Blackhawk Restaurant, moved to Morton Grove, as did the Ray Miller
Orchestra when the College Inn was closed. *Variety* claimed that the
more inexpensive roadhouses, with no cover charge, catered to the younger set more successfully than the famous in-town cabarets. Two couples
piled into a "second hand flivver with a selling price within reach of
young purses" ("a 35-buck lizzie"). With a $2.50 pint of bootleg alcohol,
they spent .66 on gas, $1.20 for twelve bottles of pop, $1.50 for two bottles
of ginger ale for a grand total of $2.93 per couple.

By May 1928, Cook County had issued over one hundred new licenses for roadhouses, thanks in part to the "wet" views of President
Anton J. Cermak of the Cook County Board of Commissioners. A
"broad-minded gent," Cermak had only two officers assigned to clear
roadhouses of vice and gambling. The others were directing traffic. In the
countryside, moreover, federal surveillance was diluted, making raids
logistically difficult. But reformers compiled a dismal composite portrait
of these roadhouses:

> ... a one- or two-story, queerly-named frame structure containing
> an improvised old-fashioned bar and a medium-sized, dimly-lit, ill-
> ventilated and smokey dining room filled with decrepit chairs and
> tables covered with soiled linen, an automatic instrument and radio
> or jazz band, a rough dance floor and decorated with twisted
> streamers of brightly-colored, gaudy crepe paper.[27]

Roadhouses never managed to provide as important a foundation for the presentation of jazz as had the inner city cabarets. Although many of them sprang up, especially in the late twenties, relatively few presented jazz. Some specialized in prostitution or gambling or drinking, leaving only a handful that presented live music for dancing. The few that offered food, drink, dancing, and entertainment, however, could generate some of the volatile emotions that surged through the best inner city cabarets. One observer who worked in roadhouses later wrote that their impact sometimes could match that of a good religious revival:

> The slow syncopation of the orchestra, the suggestive but joyful
> dancing; the low colored lights, the late hour, the drinking, and the
> general merriment of the place react powerfully on the organism of
> the individual. Both the revival and the roadhouse tend to condition
> the person to a specific way of life.[28]

By 1929, when they were investigated by the Juvenile Protective Association, many of the roadhouses supplied music either with "an automatic musical instrument," the Amplivox, the first commercially marketed juke box, or with a radio or a player piano. Whether or not a roadhouse presented a live jazz band depended on the size of the owner's investment and the ease with which substantial numbers of Chicagoans could easily travel out from the city. Oh Henry Park Dance Hall, a roadhouse fifteen miles from Chicago and easily reached on the Joliet Road or by the Joliet-Chicago interurban line, became a popular spot known for good dance music. The New Dells and the Light House in Morton Grove presented some of the top white bands because they regularly attracted wealthy upper-middle-class residents from nearby Evanston and other northern Gold Coast communities.

But the jazz age thrills of the black-and-tan cabaret experience rarely made it to the countryside. The roadhouses of Cook County during the latter part of the twenties were racially segregated. Not one of the many JPA roadhouse investigations turned up a rural black-and-tan. In a working-class Chicago Heights roadhouse, a small place which had recently been run by an African-American, the new owner proudly announced that he was "barring the dinges" and put up a sign "For Whites Only." Another JPA investigator visited a Robbins, Illinois, roadhouse in which all of the customers were black and asked the black bartender, "Do you run along the same lines as the places on the South Side of Chicago?" The bar tender replied: "No, this is pretty much of a colored place. Sometimes white folks do come in, but it is mostly for the colored people."[29]

Thus, an overpowering convergence of technological, economic, political, and social forces destroyed the unique Jazz Age cultural context that had given jazz music such particular resonance in Chicago. Beginning in 1927, therefore, many, if not all, of the jazz musicians who had made the city a famous jazz town began to move elsewhere, mainly to New York. Although the same moralizing and technological assault on jazz dominated the eastern metropolis, the political climate there closely resembled that in Chicago in 1919 when Big Bill Thompson had been elected mayor for a second term. Jimmy "Beau James" Walker, a flashy sporting set associate of Broadway promoters, actors, song writers, and underworld financiers, had been elected mayor of New York in 1926, ushering Manhattan into just the sort of political climate to encourage speakeasies and jazz. Political ties between downtown white power brokers and Harlem politicians replicated the kind of political arrangements which had tied Chicago's South Side to Thompson's City Hall.[30]

Many of the white jazz musicians who left Chicago for New York in 1927 quickly succeeded. Beginning in March 1928, Ben Pollack took his band from the Venetian Room of Chicago's Southmoor Hotel first to New York's Little Club and then to the Park Central Hotel. There, for a year and a half, Jimmy McPartland, Benny Goodman, and Bud Freeman earned some of the best salaries in the business, playing Pollack's impressive combination of technically sophisticated and emotionally hot dance music. Goodman, moreover, quickly moved into the lucrative studio scene in New York, playing jingles, popular songs, arrangements of all types of dance music, and jazz with all-white groups. He had gotten started in Chicago, but he built his swing career in New York. Although certain of the rough-and-tumble qualities of early Chicago jazz would continue to color Goodman's playing style and his bands, the clarinetist did not continue to identify himself professionally as a Chicagoan.[31]

McPartland, Freeman, and Frank Teschemacher played occasional recording sessions with the New York studio establishment, bringing what New Yorkers heard as a hotter, more seriously creative approach to musical entertainment. The white Chicagoans in New York benefited from their contacts with MCA. When the Pollack band lost its hotel job, MCA booked them into Young's Million Dollar Pier in Atlantic City— owned, of course, by Chicago's Ernie Young. Some of the Pollack men also subsequently worked in Chicagoan Paul Ash's Paramount Theater orchestra.[32]

Several of the rebel white Chicago jazzmen, under the leadership of jazz promoter and vocalist Red McKenzie, who worked for the Brunswick Record Company, also transferred their careers to New York in the late spring of 1928. Eddie Condon, Joe Sullivan, and Gene Krupa left jobs in different Chicago dance bands, expecting to work in New York backing Chicago shimmy dancer Bee Palmer in a Broadway speakeasy, but they ended up playing for another dance team in vaudeville. Though soon unemployed, they did hold several excellent recording sessions in New York which further set the foundations for the white Chicago style.33

Even though prohibition politics, radio, movie sound tracks, juke boxes, and a reeling economy destroyed much of Chicago's cabaret business after 1929, most of the dance halls survived, and some new ones were built. The larger, more prestigious dance halls had, as we have seen, already established a privileged position with Chicago's urban reformers, and therefore largely escaped the local political heat which the reformers had turned on the cabarets in 1927. In fact, as reform pressure on cabarets and speakeasies increased late in the decade, the relative innocence of the JPA-approved dance halls recommended itself to reformers and politicians. Paddy Harmon, speaking before the International Association of Policewomen in 1927, said: "You can't keep fifteen-year-olds out of the dance halls. If you do, they'll go to cabarets where they get liquor ... a fifteen-year-old girl is a problem when she's full of liquor."34

Versatile, technically skilled jazz musicians, therefore, continued to find work in the dance halls after the collapse of the cabarets. While it is true that established black dance bands of the twenties were fired beginning in 1927, this should not be interpreted as evidence of the collapse of the dance hall business in Chicago. Charles Cook's Orchestra was let go in March 1927 by Harmon's Dreamland after a four-year run, but the dance hall did not close. Harmon hired another black group—Clifford "Klarinet" King and his Orchestra—as Cook's replacement through 1928. Similarly, Jerome Pasquall, back from his studies at the New England Conservatory of Music, headed the band at Harmon's Dreamland until the crash. Cook moved first to Chicago's Municipal Pier and then to the White City Ballroom, where he stayed until the crash. Much the same can be said of the firing of Charles Elgar's orchestra from Harmon's Arcadia Ballroom. Darnell Howard and then Walter Barnes replaced Elgar, thereby keeping these (now perhaps less lucrative) jobs among all-black Local 208 musicians.35

The dance hall buildings themselves represented substantial investments which owners were reluctant to abandon. Large and small dance halls could cater to the mass market of "little guys" who could afford the fifty- or eighty-cent price of entry. *Variety* labeled dance halls "the great American playground ... for the great American peasantry ... a melting pot ... a caldron of emotions." Cabarets had been more expensive, some of them notorious for inflated cover charges, expensive drinks, and bloated tabs. As the cabaret era ended, the dance hall era began. Jazz musicians who were technically prepared could still contemplate work opportunities in the dance halls.[36]

Near the end of the decade, a chain of new dance halls owned by one company was constructed in the black neighborhoods of some of America's largest cities, finally developing a market which had been served in Chicago only by the Royal Gardens and Dusty Bottom dance halls. In 1927, as part of this chain, dance hall entrepreneur I. Jay Faggen and a group of white investors undertook to build in Chicago a state-of-the-art dance hall on the South Side. Faggen had owned, at one time or another, New York City's Roseland, the Blue Bird (after 1924 the Arcadia) Dance Hall at 53rd and Broadway, the Rosemont Ballroom in Brooklyn, Harlem's Savoy Ballroom, three Roseland ballrooms in Philadelphia, and the Cinderella Ballroom at Madison and Central streets in Chicago. He associated with white Chicago dance-band leader Ray Miller to form the Cosmopolitan Orchestras Booking and Promotion Offices to control and promote the bands serving his own and others' dance halls. The Savoy Ballroom complex, which included the Regal Theater, was opened on Thanksgiving night in 1927 at 47th Street and South Parkway (currently Martin Luther King Drive), and the Regal Theater opened on February 4, 1928. The most elegant elaborate, and expensive entertainment complex ever built in black Chicago, the Regal/Savoy pulled bright light entrepreneurs away from the Stroll and created a new center of black Chicago nightlife for the thirties.[37]

But this famous Chicago dance hall marked both the end of Chicago's jazz age and the beginning of a new era. The Savoy outshone all of the smaller South Side dance halls along the Stroll, and its success killed the 35th and South State Street entertainment district, which had been the center of Chicago's Jazz Age. (Dave Peyton tried to save the Stroll by opening the short-lived Club Congo in September 1931.) Faggen, pursuing a two-band policy at the Savoy from 1927 to 1929, opened with the well-established dance orchestras of Charles Elgar and Clarence

Black. The former was soon replaced by the Carroll Dickerson Orchestra featuring Louis Armstrong and drummer Zutty Singleton. Encouraged by the Savoy Ballroom job, the Dickerson orchestra became the hottest dance band on the South Side and in 1929 broadcast nightly over WMAQ, the Chicago *Daily News* affiliate. Young dancers began to create "wild new dance steps ... the Bump, Mess Around, and the Fish Tail ... dislocating the joints of the younger generation."[38]

But I. Jay Faggen refused to limit hiring to local South Side or even Chicago bands. He regularly booked white bands from Chicago and New York into the Savoy, most notably the theater stage bands of Paul Ash and Benny Meroff, stars of the Loop's largest theaters. Black band-leaders predicted doom from this invasion of their territory. Faggen spearheaded the quickly accelerating movement toward large, versatile theater stage/ballroom orchestras by installing New York's Fess Williams and His Jazz Joy Boys, who had long appeared at his Harlem Savoy Ballroom, into Chicago's Regal Theater and Savoy Ballroom.[39]

Amid changing economic, technological, and organizational conditions which left so many of Chicago's cabaret musicians unemployed and disheartened, first Louis Armstrong and then Earl Hines became leaders and symbolic heroes of Chicago's struggling jazz musicians of the twenties. Armstrong in particular adapted his cabaret experiences to the new outlines of musical show business. He triumphed in each of the new formats created near the end of the decade. First, of course, the trumpeter and vocalist became the star attraction in the newest, largest, and best connected dance hall in black America, playing to packed houses and rave notices with the Dickerson band at the Savoy. Armstrong was the hero there one night in April 1928 when owner I. Jay Faggen took the musicians by surprise. Inviting the Erskine Tate Orchestra to appear at the Savoy in addition to the regular Dickerson and Clarence Black bands, Faggen got the idea of having all three bands play a number named after his ballroom:

> A note was handed to the house orchestras by Mr. Fagin, asking them to join in and play "Savoy Blues" with the guest orchestra. This was wonderful. Three bands, consisting of 37 players, rocked the beautiful ballroom with their scintillating music. It was Louis Armstrong, a member of Carroll Dickerson's Orchestra, whom this writer has termed the "Jazz Master," who saved the hour. As the

boys would say, Louie poured plenty of oil and it soaked in too. The crowd gathered around him and wildly cheered for more and more.[40]

South Side musicians also could count on Armstrong's reassuring ability to outshine the top white solo instrumentalists in bands that were being brought into the Savoy to play opposite the Dickerson band. Dave Peyton gloated that the cornet star "has slaughtered all the ofay jazz demons appearing at the Savoy recently." Peyton began referring to Armstrong as "King Menelik," after the nineteenth-century Ethiopian emperor who drove the Italians out of Ethiopia. Armstrong enjoyed the added prestige of appearing on stage with Peyton's orchestra at the Regal Theater, demonstrating his ability to adapt to the new stage band strategies and vastly reinforcing the popularity of a very expensive orchestra in bad economic times.

In 1929, Armstrong worked through I. Jay Faggen to arrange a two-night feature engagement with the Luis Russell band in Faggen's Harlem Savoy Ballroom. The trip to New York included recording sessions for Okeh. Armstrong, Faggen, and Okeh executive Tommy Rockwell subsequently planned a second Harlem appearance for Armstrong two months later in May. The trumpeter, by now aware of his national status as a star attraction, brought the Dickerson band, which had done so much for him at Chicago's Savoy Ballroom, along with him on that second trip to New York presenting Rockwell with a *fait accompli*. When Armstrong's Savoy appearances and a long run in the Broadway show *Hot Chocolates* ended, Rockwell refused to promote the Dickerson musicians, most of whom returned to Chicago. But Armstrong had more than proved his loyalty to a fine group of musicians.

Armstrong's success in *Hot Chocolates* confirmed the original ambitions of dozens of black cabaret entertainers who had hoped to move from the black-and-tans into the legitimate theaters. Peyton chortled that Armstrong made "Broadway eat out of the palm of his hand."[41]

Earl Hines, the other great symbolic hero of South Side jazz lovers, carried Chicago's cabaret traditions to success in the 1930s. Hines toured the east coast cities, playing dance halls and white and black theaters as a stage band attraction. In Chicago, his band appeared at the Savoy Ballroom as well as the Grand Terrace, where Ed Fox arranged for his band to back New York dancers such as Maude Russell, Bill "Bojangles" Robinson, and Buck and Bubbles.[42]

By 1932, a growing trend toward large jazz-oriented black dance bands was clearly visible in Chicago. Earl Hines, of course, had been building his band at the Grand Terrace since the end of the previous decade, keeping an edge on the competing black big bands led by Tiny Parham, Reuben Reeves, Walter Barnes. Black Chicago also welcomed big bands from Kansas City and New York as well, cheering the Benny Moten, Duke Ellington, and Fletcher Henderson orchestras on their periodic tours through the South Side. According to the *Defender,* all of the most ambitious South Side Chicago band leaders, including Jimmie Noone and Eddie South, vied for national exposure, eagerly seeking the promotional backing necessary for national tours like the one which Ed Fox promoted for Earl Hines in 1932. Touring bands now played a maximum of a week or two in large Chicago cabarets, hotels, and dance halls before moving on.[43]

Some white jazzmen struggled under often discouraging conditions in the clubs after the stock market crash. Art Hodes stayed on in Chicago until 1938, long after many other white players had left for New York. After the crash, Hodes survived by playing in Floyd Town's and Frank Snyder's dance bands and playing solo piano at the Capitol Dancing School, Harry's New York Bar, and the Liberty Inn. His description of the last underscores how, under adverse economic circumstances, jazz returned to a lesser, supporting role in cabaret entertainment.

Johnny McGovern owned the Liberty Inn at Clark and Erie and the connecting hotel next door. He featured bar girls in the front and vaudeville-style cabaret acts and strippers in the back room. Hodes found it "a grind" playing for the floor show and dancing, but a number of important white musicians—pianist Bob Zurke, trumpeters Marty Marsala, Wingy Manone, and Louis Prima, and drummer Earl Wiley—played with him in the backroom of the Liberty Inn at one time or another. Sam Beer's My Cellar on State near Lake Street kept jazz and racial integration alive by featuring inter-racial jam sessions through the repeal of prohibition.[44]

Most of the rebel white Chicagoans, whose enthusiasm had done so much to enliven the Chicago jazz scene, migrated to New York and survived during the lean years between 1929 and 1935 by playing in various dance bands—Arnold Johnson, Roger Wolfe Kahn, Frank Westphal, Meyer Davis, Jan Garber, and others. Some of these dance band leaders had worked extensively in Chicago during the 1920s, and had preceded the jazzmen to New York. Often booked by MCA, these bands continued to play an important role in the professional and artistic lives of the white Chicago jazzmen.[45]

As the depression waned and World War II approached, however, these musicians began a long process of appropriation of Chicago's jazz band tradition. "Chicago Jazz" became closely associated with them, although as we have seen, they had actually performed relatively little in Chicago during the 1920s. Unlike Louis Armstrong, who adapted his cabaret experiences to the theater stage band in a pragmatic approach to musical show business, the immigrant white rebel Chicagoans closely connected their musical aspirations to the remembered sensibilities of twenties Chicago and its cabaret/speakeasy tradition. Red McKenzie and Eddie Condon had attached the city's name to their first records in 1927 (and Condon would continue to do so in several subsequent recording sessions); *Variety* touted their first "Chicagoans" records as the latest thing soon after their arrival in New York. Accessible to journalists, Condon, McPartland, Freeman, Wettling, and others publicized the importance in their lives of Chicago, Austin High School, and Windy City gangland speakeasies.[46]

In the mid-thirties, as the Swing Era began to dominate the national scene, a Chicago-influenced small jazzband style began to emerge in New York. As early as 1932, Milt Gabler, proprietor of the Commodore Music Shop at 144 East 42nd Street began reissuing older jazz records by both black and white groups. A small but enthusiastic market existed for such reissues among white professionals like English professor Marshall Stearns, *Time-Life* illustrator Richard Edes Harrison, commercial artist Paul Smith, *Time* and *Fortune* staff writer Wilder Hobson, and a very young George Avakian, who later came to be in charge of popular music albums at Columbia records.[47]

Reissue programs and their accompanying brochures and liner notes played an influential role in a retrospective definition of "Chicago Jazz." Stearns organized a group called the United Hot Clubs of America, devotees of small band jazz music as an art form, who began an influential reissue series under their own label. To promote a taste for small band jazz in the midst of the Swing Era, Gabler organized in 1938 a series of insider, invitation-only jam sessions at the Decca studios at 799 Seventh Avenue at the corner of 52nd Street.

Beginning in 1938, several of the transplanted rebel Chicagoans began playing regularly at a club called Nick's, on Seventh Avenue and

Tenth Street in Greenwich Village, which was frequented by journalists, artists, and writers. While black solo stars such as Sidney Bechet and Willie the Lion Smith were featured from time to time, Nick's usually presented white bands to white audiences. In the process, "Chicago Jazz" was even more closely associated with the rebel white Chicago jazzmen. At about the same time, when Milt Gabler began recording both black and white jazz groups, several of the white ones appeared under names associated with Chicago.[48]

George Avakian solidified the identification of Chicago jazz with some of the white musicians who had played jazz there in the twenties when he produced and wrote the liner notes for "Chicago Jazz," the first jazz album ever made, which featured white bands led by Eddie Condon, Jimmy McPartland, and George Wettling and was issued on Decca. Avakian later directed several recording sessions at Columbia Records featuring "Chicago" musicians, like Wettling, Bud Freeman, and Bill Davison, all of whom were closely associated with Eddie Condon.[49]

Condon's influence in defining "Chicago Jazz" grew, thanks in great part to a circle of influential advertising executives that included Ernest Anderson and Condon's wife Phyllis Smith. He began to front a series of jazz concerts at Town Hall, which lasted for several years. Condon industriously promoted the Chicago traditions as he understood them. His autobiography *We Called It Music* further emphasized the importance of the speakeasies to white Chicago jazzmen during the twenties while also emphasizing the importance of the South Side jazz scene. He opened a long-lived jazz club called "Condon's" in Greenwich Village in 1945 and featured most of the white Chicagoans at one time or another playing very loosely organized jazzband music. When, during and after World War II, the traditional jazz revival flourished, Chicago jazz came to be exclusively associated with solo-dominated dixieland played by white musicians. In the process, much of the interesting variety of the music that had been played and recorded in Chicago during the twenties was lost.[50]

The earliest jazz writers, working from phonograph records and oral interviews with the musicians, further underscored the importance of a "Chicago style" jazz played by white musicians. French critic Hugues Panassié's influential book *Hot Jazz* was published in America in 1934. He was deeply impressed by the McKenzie and Condon's Chicagoans records; he defined Chicago style jazz as unarranged, small group im-

provisation characterized by a minimalist, few-notes style and a preference for hot intonation and intensity of feeling over virtuosity. Panassié
insisted that the style, which peaked between 1925 and 1929, was a "white
appropriation" of Louis Armstrong's playing style.[51]

Wilder Hobson, the first American to write a book of jazz criticism,
defined the Chicago style as a blend of "the Negroes' personal intensity
and a linear economy suggestive of Beiderbecke." Hobson believed that
Chicago style gained focus when compared with New York cornetist Red
Nichols's and trombonist Miff Mole's 1929 "white manner of playing" in
New York. Their "clean, agile, polished, graceful, and balanced" style
contrasted with the wilder more expressive spirit which the white Chicagoans had assimilated from black musicians. Hobson also wrote that
Chicago style stood out in bolder relief when compared with the "blatant
virtuoso exhibitions of the 'swing' fad."[52]

On the other hand, at the end of World War II, writer Rudi Blesh
began a strong critical counter-reaction against the soaring reputation of
the rebel white Chicagoans. While convinced that it was "impossible to
praise too highly the ardor and the fidelity of their pursuit of jazz," he
said the "Chicagoans" played only a "white imitation of Negro jazz,
sincere but not profound," demonstrating a progressive loss of control
and a consolidation of the weakest features of South Side jazz. These
"white faced boys," Blesh wrote, played a "bad, good music," while all
but a few "faithful Negroes," "weaned away by white commercialism
and the easy success of their own ruined music, no longer wish to play
jazz."

Blesh and other jazz writers believed that the popularity and reputation of the white Chicagoans was belied by their comparatively thin
contributions to jazz in the twenties. And musicologist Gunther Schuller,
writing in 1968, buried the rebel Chicagoans in a footnote that labeled
their records as "commercial performances geared to a thriving mass
market requiring a consumer's product." Jazz writer Martin Williams,
Schuller's frequent collaborator, omitted most of the white Chicagoans
from his collections of essays and from his vastly influential record set the
"Smithsonian Collection of Classic Jazz."[53]

A cultural history of jazz in Chicago during the twenties clearly
demonstrates that the label "Chicago Jazz" has been far too musically
and racially narrow a term to adequately describe the variety of approaches to jazz actually played and recorded there in the 1920s. As a
stylistic label, the phrase has suggested the spirit of fast-paced, nervous

excitement that Chicago stimulated in musical entertainment; but, as we have seen musicians manipulated this complex set of sensibilities in many unexpected directions. The white movement to appropriate Chicago jazz, while paying lip service to the priority of the South Side Chicago jazz scene, removed the music from the context of black cabaret show business in which it had grown, completely altering its historical and cultural context.

But this historical investigation of jazz activity in Chicago during its hey day there in the 1920s documents several more functional dimensions of jazz, not as a particular stylistic approach to making music, but as a cultural force which interacted with some of the most important trends in northern urban society after World War I. The music of King Oliver, Louis Armstrong, Jelly Roll Morton, Bix Beiderbecke, and Frank Teschemacher expressed and stimulated nonverbal patterns of emotional excitement which Chicago stirred in its youth. Jazz in Chicago gave voice to a sense of abandon in the new urban wilderness.

And just as importantly, jazz functioned to express and channel some of the explosive emotions generated by racial and cultural tensions in Chicago. The opening bars of Chicago's post-World War I jazz age music unfolded in South Side musical institutions in 1918–19 at the time of the city's worst race riot. Chicago's Roaring Twenties were born in an atmosphere of racial hatred and fear; the black-and-tan cabarets, whose owners often provided invaluable political organization and recruitment for Big Bill Thompson's Republicans, functioned to provide channels for the resolution of racial tensions through intense, structured rituals of nonverbal inter-racial expressiveness. As Victor Turner has explained,[54] conflict, an integral part of the social process, produces "social dramas," incidents like Chicago's terrifying race riot of 1919. After an emotional, bloody social conflict, some redressive action, institutionalized or *ad hoc,* is usually brought into action by representative members of the disturbed social system.

The process of reintegration of society, never complete, usually involves an invitation of members of the different parties to a major ritual which will affirm the other side of their conflictual relations—*communitas*—nonrational (but, as Chapter Five argued, not necessarily "irrational") bonds uniting people over and above any formal social associations which may unite them. An intuitive, "liminal" communion, labeled "liminoid" when Turner writes of its night-club variant, emerged through drinking illegal alcoholic beverages, dancing, and listening to jazz in Chicago's cabarets. Such liminoid rituals sometimes produced an emotional catharsis, such as that experienced by Mezz Mezzrow and analysed in Chapter Four, marking an exchange of qualities between the groups in conflict, and causing, in some instances, genuine transformations of character and social relationships. Relationships between members of formerly conflicting racial groups, Louis Armstrong and Wild Bill Davison or Mezz Mezzrow, for example, become antistructural, egalitarian, direct, nonrational, existential ones rooted in the shared experience of music, movement, nomadism, and transience of the jazz musician's life.

Liminoid experiences, such as those produced by jazz in twenties Chicago, were often interpreted as sacred by those who experienced them, forming a bedrock change in sensibilities. These moving experiences of African-American music and dance, moreover, remained locked in memory, capable of eliciting, as they do in the many autobiographies of white jazz musicians, strong emotions over the passage of many years. In this way, nearly fanatical loyalties arose to "Chicago Jazz" or to Louis Armstrong's Hot Five and Hot Seven records, or to Bix Beiderbecke, among people who not only loved the music but also shared some sense of racial conflict or an awareness of social injustice in America. Armstrong or Beiderbecke or Jimmie Noone often became shamanistic figures, their very musical instruments potent symbols of their magical powers, their records sacred texts, and their photographic images powerful icons.

So, too, prohibition, the immediate triumph and ultimate failure of the Victorian reform movement, added power to the twenties' jazz experience in Chicago. The leading cabaret customers included undercover F.B.I. agents, plainclothes city policemen, note-taking observers from the Juvenile Protective Association, leading city politicians, bootleggers, and gangsters in a volatile mixture that added an extra jolt to the incendiary music and the burning bootleg alcohol. This cultural context heightened the meaning of the experience of jazz in twenties Chicago.

Finally, the unusual mixing of social classes that came to be associated with jazz audiences stimulated its own special thrills. The admixtures of slumming Gold Coasters with tourists, young office workers, pimps, gangsters, prostitutes, politicians, and government agents created an unprecedented fluidity and indeterminacy of social status. For an evening, at least, rich, workaday, and poor could ogle one another, aping each other's style and gestures. Whites could mimic black music and dance; blacks could dress up in tuxedoes and gowns and dance and drink with whites; the poor could revel in expensive leisure time elegance; the rich could have fun acting like tough street lords; rural tourists could rub shoulders with urbane sophisticates.

None of this level of meaning, however, can present itself to conscious understanding when jazz history is built only on the phonograph records, career summaries of individual musicians, and photograph collections. Seen in that familiar perspective, jazz has appeared to be a musical art form evolving in its own isolated world of instrumental mastery, chord progressions, and orchestral formations and disintegrations. These chapters have tried to restore some sense of the cultural context in which jazz functioned historically, and to describe the racial and cultural forces which it expressed at a given time in a specific place. Jazz in Chicago during the Roaring Twenties gave voice to and commented upon the intensely felt emotions generated by the excitement and dangers of early twentieth-century urban life.

Appendix

THE FOLLOWING LISTS summarize the chronology of recording activity either in Chicago itself or elsewhere of groups closely associated with the midwestern city. They should be seen as suggestive and illustrative rather than definitive. In some cases, record collectors, music scholars, and jazz writers disagree about which records do and do not present jazz music. Although most jazz criticism has exmphasized African-American jazz, some now argue that many more records by white dance bands should be included in jazz discographies; others would disagree. Until such time as scholarship produces more definitive discographies of jazz records, I have followed the distinctions established by discographer Brian Rust between jazz and dance band records. Although Rust remained vague about his reasons for including and excluding records from his discographies, the data I gathered on the different cultural contexts within which jazz and dance groups functioned support his basic distinctions between jazz records and "American Dance Music" records. Therefore I have included as "jazz records" those sides by major white dance orchestras included by Rust in *Jazz Records*. The records by white dance orchestras that Rust excluded from his jazz discography have been similarly omitted from list A.

Ambiguities also cloud a few decisions about whether or not a lesser known recording group deserves to be included as a Chicago band. These lists focus upon instrumental recordings made in Chicago, Richmond, Ind., Camden, N. J., Grafton, Wisc., and New York City before 1931 by musicians with some association with Chicago beyond the occasional or isolated Chicago recording session. When in doubt about some of the more obscure groups, however, I have usually included them.

In order to retain the focus on groups involved in Chicago night life, the many records made in Chicago by groups like the Detroit-based McKinney's Cotton Pickers have been excluded because neither the individual musicians nor the orchestra worked extensively in Chicago clubs or dance halls during the 1920s and because such orchestras recorded more often in other cities. So too, when a leader such as Jelly Roll Morton moved to New York and continued to record, but with a substantially different group of musicians, I have omitted those New York records in spite of the fact that they appeared under the same Morton Red Hot Peppers trademark. If, as in the case of several bands, important Chicago musicians remained in what otherwise became a New York recording band, I have included such recordings. The same desire to highlight Chicago nightlife led to the inclusion of the Coon-Sanders Original Nighthawk Orchestra that started in Kansas City and employed Kansas City musicians but moved to Chicago for lengthy jobs at the Congress Hotel and Blackhawk Cafe.

Jazz recording chronology can be traced through the first date in the left-hand column. In each of the two lists, that date represents the first recording session by a given group that produced jazz as defined by discographer Brian Rust. The second date in the left-hand column indicates the last "jazz" recording session by that same group in the period considered in this book.

The right-hand column indicates the number of jazz sides (individual recorded performances) made by each group and the company labels involved. Repeated takes of the same number are included, but rejected takes are excluded. The abbreviations of recording company labels are as follows:

Vic	= Victor Recording Company;	Ok	= Okeh;
Bp	= Black Patti;	Oly	= Olympic;
Bwy	= Broadway;	Od	= Odeon;
Br	= Brunswick;	Pm	= Paramount;
Chg	= Challenge;	Auto	= Autograph;

Col	= Columbia;	Cx	= Claxtonola;
Ed	= Edison Diamond Disc;	Sil	= Silvertone (American);
Gnt	= Gennett;	Spr	= Superior;
Her	= Herwin (American);	Spt	= Supertone (American);
Hom	= Homochord;	Sw	= Swaggie;
Mt	= Melotone (American);	Uhca	= United Hot Clubs of America.

Despite the hazy musical distinctions between jazz and dance band music, taken together these lists survey the general order in which jazz groups recorded in Chicago. The first and last dates of recording activity suggest the relative length of time during which any given group was considered ready to record jazz oriented material. One can also see which ones were most active in the field as well as their relatively longer or shorter periods of recording activity.

A. CHICAGO JAZZ RECORDS: WHITE BANDS

First—Last Recording Dates	*Recording Group*	*No. Sides & Labels*
4-11-21—5-23-25	The Benson Orch.	34 Vic
8-?-21	Lindsay McPhail	2 Oly
2-?-22—5-15-30	Isham Jones Orch.	32 Br
3-9-22—3-10-22	O'Hare's Super Orchestra	6 Gnt, Voc
6-?-22—9-23-26	Russo & Fiorito's Oriole Orchestra	22 Br, Vic
8-18-22—3-6-24	Frank Westphal Orchestra	16 Col
8-29-22—3-26-25	New Orleans Rhythm Kings[1]	33 Gnt
8-31-22—3-18-24	Roy Bargy	6 Vic
3-?-23—5-?-23	Albert E. Short's Tivoli Syncopators	5 Voc
4-2-23	Original Memphis Melody Boys	4 Gnt
4-?-23	Guyon's Paradise Orchestra	3 Ok
5-30-23	Chicago Blues Dance Orchestra	2 Col
6-1-23—7-13-23	Art Landry & the Call of the North	9 Gnt
6-?-23—8-25-28	Charley Straight and Orchestra	44 Pm, Br

First—Last Recording Dates	Recording Group	No. Sides & Labels
7–10–23—11–12–23	Porter's Blue Devils	10 Gnt
8–?–23—9–13–28	Paul Ash Orchestra	12 Br, Col
10–16–23—6–27–27	Art Kahn Orchestra	14 Col
10–18–23—12–5–23	Midway Garden Dance Orchestra	7 Col, Pm
12–?–23—1–?–26	Joie Lichter's Strand Symphonists	10 Pm, Gnt
2–18–24—12–12–24	Wolverine Orch. (Beiderbecke)	18 Gnt
2–23–24—11–14–29	Mound City Blue Blowers	18 Br, Col Vic
2–25–24	Bucktown Five	7 Gnt
3–10–24—11–27–28	Guy Lombardo Royal Canadians	11 Gnt, Col
4–3–24	Paul Biese Orch.	3 Vic
4–5–24—12–12–29	Coon-Sanders Original Nighthawk Orchestra	66 Vic
4–?–24—c.5–?–24	Boyd Senter	10 Auto, Pa
7–?–24	Chicago DeLuxe Orchestra	2 Auto
9–?–24—3–2–26	Merritt Brunies Friars Inn Orch.	15 Auto, Ok
10–?–24—6–26–26	Al Turk	10 Oly, Ok
11–?–24—1–?–25	Dell Lampe Orch.	8 Auto
11–7–24—12–17–25	Jack Chapman Drake Hotel Orchestra	3 Vic
11–7–24—12–16–25	Ralph Williams Rainbo Orchestra	7 Vic
12–5–24	Frankie Quartell Orchestra	4 Ok, Br
1–26–25—9–8–30	Bix Beiderbecke Groups	20 Gnt, Ok Vic
7–?–28—12–?–28	Bill Haid Groups	25 Pm, Auto
3–?–25—7–?–25	Super Syncopators	4 Auto
4–30–25—12–17–25	Fred Hamm Orchestra	6 Vic, Br
5–?–25	Voltaire De Faut	1 Auto
6–?–25	Dixie Boys	2 Auto
c.7–?–25	Stomp Six	2 Auto
8–3–25—11–14–27	Don Bestor Orch.	5 Vic
8–8–25—1–?–30	Benny Meroff Orch.	4 Ok, Br
11–23–25	Bob Deikman Orch.	3 Gnt
12–9–26—6–23–30	Ben Pollack Orch.	52 Vic
5–12–27—12–15–27	Sol Wagner Orch.	5 Ok
7–19–27	Bernie Schultz Cresent Orch.	4 Gnt, Spr, Chg
10–2–27—12–21–29	Ray Miller Orch.	18 Br

First—Last Recording Dates	Recording Group	No. Sides & Labels
10–?–27	Chicago Loopers	3 Pa
10–12–27—3–25–29	Original Wolverines	17 Br, Voc
10–28–27—11–20–30	Hoagy Carmichael	14 Gnt
11–11–27—1–9–28	Emil Seidel Orch.	21 Gnt, Sil, Spr
12–8–27—12–16–27	McKenzie & Condon's Chicagoans	4 Ok
1–10–28	Husk O'Hare's Wolverines	2 Voc
1–23–28—8–13–29	Benny Goodman Groups	10 Br, Voc
1–24–28—1–27–30	Louisiana Rhythm Kings	31 Voc, Br
2–?–28—4–?–28	Charles Pierce & Orchestra	7 Pm
3–20–28	Tracy Brown's Orch.	5 Col
3–29–28—9–27–28	Thelma Terry's Playboys	6 Col
4–4–28—4–28–28	Chicago Rythm Kings /Jungle Kings	6 Br, Voc, Pm
4–?–28—11–9–28	Arnold Johnson Orchestra	5 Br
4–28–28	Frank Teschemacher's Chicagoans	1 Uhca
6–?–28	Billy Stennett's Carolina Stompers	4 Bwy
7–5–28	Frankie Masters Orch.	1 Vic
7–28–28—2–8–29	Eddie Condon Groups	8 Par, Ok, Vic
7–?–28—8–?–28	Midnight Serenaders	2 Voc
9–4–28—9–19–30	Wingy Manone Groups	10 Ok, Col, Voc
10–22–28	Danny Altier Orch.	2 Voc
11–16–28—11–17–28	Lawrence Welk	4 Gnt
12–3–28	Bud Freeman & Orch.	2 Ok
4–22–29	Johnny Burris Orch.	2 Gnt
4–26–29	G. Belshaw KFAB Orch.	2 Br
7–?–29	Nobby Neal & Al Lyons	2 Pm
10–2–29	Syd Valentine Patent Leather Kids	4 Gnt
10–18–29	Elmer Schoebel & Friars Society Orch.	2 Br
2–12–29—4–11–30	Frank Melrose Groups	12 Pm, Gnt
1–24–30	The Cellar Boys	3 Voc
2–?–30	Louis Panico Orch.	2 Br

[1] This group first recorded on August 29, 1922, under the name "Friars Society Orchestra." A very similar, if smaller, group of musicians subsequently recorded as the "New Orleans Rhythm Kings" beginning on March 12, 1923. Jelly Roll Morton played on five recordings (three different tunes, five takes) at NORK's July 17, 1923, recording session so that one session qualifies as "racially integrated"; whites otherwise dominated the group's personnel.

B. CHICAGO RACE RECORDS: JAZZ[2]

First—Last Recording Dates	Recording Group	No. of Sides & Labels
1–?–23—12–?–28	Clarence M. Jones	9 Auto, Pm
4–6–23—4–27–27	King Oliver Bands	66 Gnt, Col, Ok, Voc
6–1–23—2–8–29	Richard M. Jones Jazz Wizzards	26 Ok, Gnt, Voc
6–?–23—6–10–27	Jelly Roll Morton Groups	70 Gnt, Vic, Pm, Ok, Voc
6–23–23—5–28–26	Erskine Tate Vendome Orch.	4 Ok, Voc
9–?–23—1–14–25	Ollie Powers Harmony Syncopators	8 Cx, Pm
10–3–23—11–6–23	Lois Deppe	6 Gnt
10–?–23—11–?–23	Young's Creole Jazz Band	4 Pm
11–?–23—8–?–24	Sammy Williams	4 Auto
12–?–23—10–10–28	Jimmy Wade Orch.	7 Pm, Gnt, Voc
1–21–24—3–30–28	Doc Cook Groups	21 Gnt, Col, Ok
4–?–24—4–1–28	Jimmy Blythe Groups	27 Gnt, Voc, Pm
8–?–24—10–4–28	Sammy Stewart Orch.	7 Pm, Voc
11–?–24—1–?–26	Jimmie O'Bryant	27 Pm
11–?–24—8–?–26	Lovie Austin Blues Serenaders	16 Pm
2–22–25—6–?–25	Hersal Thomas	4 Ok
2–?–25	Original Midnight Ramblers	2 Auto
4–6–25—2–24–26	Hociel Thomas	15 Gnt, Ok
5–?–25	Jones' Paramount Charleston Four	2 Pm
11–12–25—12–12–28	Louis Armstrong Groups	63 Ok, Col
2–?–26	Austin's Musical Ambassadors	1 Pm
2–?–26	Chicago Hottentots	2 Voc
3–10–26—11–17–26	Luis Russell Bands	6 Voc, Ok
5–28–26	'Lill's Hot Shots	3 Voc
5–29–26—4–25–29	Jimmy Bertrand's Washboard Wizzards	12 Voc
6–21–26	Art Sims Creole Roof Orchestra	3 Ok
6–25–26—10–9–28	Albert Wynn Groups	6 Ok, Voc
7–13–26—7–14–26	New Orleans Wanderers/Bootblacks	8 Col
7–25–26	Birmingham Bluetette	1 Her
8–?–26—10–24–29	Junie Cobb Groups	9 Pm, Voc

First—Last *Recording Dates*	*Recording Group*	No. of Sides *& Labels*
8–?–26—8–28–29	Vance Dixon Groups	7 Pm
9–?–26	Preston Jackson Uptown Band	4 Pm
9–?–26	Keppard's Jazz Cardinals	3 Pm
9–?–26	Elgar's Creole Orchestra	6 Voc, Br
10–?–26	Wilson's TOBA Band	2 Pm
12–?–26	Pickett-Parham Apollo Syncopators	2 Pm
12–10–26—6–7–27	Dixieland Jug Blowers	24 Vic
1–?–27—6–29–28	Jasper Taylor Groups	4 Pm, Voc
1–?–27—6–22–29	Charles "Cow Cow" Davenport	25 Pm, Voc, Br, Gnt
3–30–27	Vicksburg Blowers	4 Gnt, Ch
3–?–27—2–7–29	Johnny Dodds Groups	40 Pm, Vic
6–7–27—2–6–29	Clifford Hayes' Louisville Stompers	19 Vic
6–?–27—9–?–27	Nelson's Paramount Serenaders	4 Pm
7–19–27—7–24–27	(King) Brady's Clarinet Band	2 Gnt
7–20–27	Hightower's Night Hawks	2 Bp
8–12–27—7–19–28	State Street Ramblers[3]	22 Gnt, Ch
8–?–27—12–?–27	Dixie-Land Thumpers	5 Pm
9–?–27—9–20–29	Will Ezell	14 Pm
12–2–27—3–12–29	Eddie South Alabamians	7 Vic, Hmv
12–?–27—11–11–30	Tiny Parham Groups	46 Vic, Pm
12–?–27	Axel Christensen	2 Pm
12–?–27	Meade Lux Lewis	1 Pm
12–3–27—7–4–28	Chicago Footwarmers	10 Ok
1–21–28	Levee Serenaders	2 Voc
4–1–28—6–7–29	W.E. Burton	6 Pm, Gnt
4–3–28	Fess Williams Joy Boys	2 Voc
5–16–28—10–30–30	Jimmie Noone Apex Club Orchestra	63 Voc
5–25–28—7–5–28	Carroll Dickerson Groups	4 Br, Od
6–?–28	Dixie Four	4 Pm
6–?–28	Tub Jug Washboard Band	3 Pm
7–?–28	Clarence Black Savoy Trio	2 Pm
9–22–28	Lil Hardaway's Orchestra	1 Voc
10–9–28	Midnight Rounders	2 Voc
12–8–28—10–25–29	Earl Hines	25 Ok, Vic

First—Last Recording Dates	Recording Group	No. Sides & Labels
12–10–28	E.C. Cobb & His Corn Eaters	2 Vic
12–12–28—7–25–29	Walter Barnes' Royal Creolians	8 Br
12–?–28	Beverly Synocopators	1 Pm
1–29–29—8–22–29	Jabbo Smith Rhythm Aces	19 Br
1–4–29—3–8–29	Ikey Robinson Bands	9 Br
2–?–29	Paramount Pickers	2 Pm
2–12–29	King Mutt & His Tennessee Thumpers	7 Gnt, Ch Spt
3–11–29	Thomas' Devils	2 Br
3–21–29	Bill Johnson's Louisiana Jug Band	2 Br
3–30–29—2–8–30	Alex Hill	7 Voc
4–?–29	Windy Rhythm Kings[4]	2 Pm
4–15–29—7–24–29	Tampa Red [Jug Bands]	9 Voc
5–9–29	Ike Rogers Biddle Street Boys	1 Br
5–22–29—10–1–29	Reuben Reeves Orch.	15 Voc
7–2–29—3–15–29	Kansas City Stompers	4 Br
7–24–29	Beale Street Washboard Band	3 Voc
8–21–29—9–11–29	Omer Simeon	2 Br
8–28–29	Harlem House Rent Stompers	1 Br
9–23–29	Dixie Rhythm Kings	4 Br
10–28–29	Kentucky Jazz Babies	2 Vic
10–?–29	Charles Avery	1 Pm
12–20–29—2–8–30	Alex Hill Groups	5 Voc, Sw
12–?–29	Bob Call	1 Br
2–4–30—11–?–30	Bob Robinson	4 Ch, Pm
2–?–30	Wesley Wallace	2 Pm
5–15–30—11–19–30	Harry Dial Blusicians	6 Voc
7–?–30—11–20–30	Lloyd Smith's Gut-Bucketeers	5 Voc
11–12–30	Frankie Franko Louisianians	2 Mt

[2] This list surveys instrumental music and does not include the hundreds of vocal blues records which more properly belong to the closely related but still separate history of the blues in Chicago.

[3] An "unknown white teenager" played alto saxophone on the July 18, 1928, recording session by this group, so that six of their sides were "racially integrated."

[4] This black group led by the Cobb brothers may have included Frank Melrose on piano and thus would qualify as a racially integrated session.

Notes

Abbreviations of Archival Collections

CHS   Chicago Historical Society

CJA   Chicago Jazz Archive, Music Division,
Regenstein Library, University of Chicago

COHC   Columbia University Oral History Collection

HCWB   Vivian G. Harsh Collection of Afro-American Literature and
History, Carter G. Woodson Branch, Chicago Public Library

IJS   Institute of Jazz Studies, Rutgers University–Newark, New Jersey

JSC   John Steiner Collection, Milwaukee, Wisconsin

SPCPL   Special Collections, Chicago Public Library

JPAR   Juvenile Protective Association Records, University Library,
University of Illinois at Chicago

WRHJA   William Ransom Hogan Jazz Archive, Tulane University, New
Orleans

Introduction

1. Gunther Schuller, *Early Jazz: Its Roots and Musical Development* (New York: Oxford Univ. Press, 1968); Martin T. Williams, *The Jazz Tradition* (New York: Oxford Univ. Press, 1970).

2. LeRoi Jones (Amiri Baraka), *Blues People* (New York: Morrow, 1963); Lawrence Levine, *Black Culture and Black Consciousness* (New York: Oxford Univ. Press, 1977); Neil Leonard, *Jazz and the White Americans* (Chicago: Univ. of Chicago Press, 1962); Chadwick Hansen, "Social Influences on Jazz Style: Chicago, 1920–1930," *American Quarterly* 12 (1960), 493–507. Kathy Ogren, *The Jazz Revolution: Twenties America and the Meaning of Jazz* (New York: Oxford Univ. Press, 1989).

3. Clifford Geertz, *The Interpretation of Cultures* (New York: Basic Books, 1973), chaps. 1, 15. *Jazz City: The Impact of Our Cities on the Development of Jazz* (Englewood Cliffs, N.J.: Prentice-Hall, 1978). Also see Janet Wolff, "Forward: The Ideology of Autonomous Art," in *Music and Society: The Politics of Composition, Performance and Reception,* ed. Richard Leppert and Susan McClary (Cambridge, Eng.: Cambridge University Press, 1987), 1–12.

4. Chicago *Defender,* April 17, 1926, p. 6.

5. Joachin E. Berendt, *The Jazz Book: From Ragtime to Fusion and Beyond,* trans. H. and B. Bredigkeit with Dan Morgenstern, new ed. (Westport, Conn.: L. Hill, 1982), 371–77. Edward Pessen explores the dangers of reverse racism in his review of Gunther Schuller, *The Swing Era: The Development of Jazz, 1930–1945* (New York: Oxford Univ. Press, 1989) in "A Less than Definitive Nonhistorical Account of the Swing Era," *Reviews in American History* 17 (1989), 599–607. For extended discussions of the proper definition of "jazz," see Mark Gridley, Robert Maxham, and Robert Hoff, "Three Approaches to Defining Jazz," *Music Quarterly* 73 (1989), 513–31, and Lee B. Brown, "The Theory of Jazz Music 'It Don't Mean a Thing . . .'," *Journal of Aesthetics and Art Criticism* 49 (1991), 115–27. The Dec. 9, 1906, *Tribune* article on opportunities for blacks in music was reprinted in the Chicago *Broad Ax,* Dec. 15, 1906, p. 2.

6. For a theoretical discussion of the origins of cultural history, see Roger Chartier, "Intellectual History and the History of *Mentalités:* A Dual Reevaluation," in Chartier, *Cultural History: Between Practices and Representations* (Ithaca: Cornell Univ. Press, 1988), chap. 1; and Lynn Hunt, "Introduction: History, Culture, and Text," in *The New Cultural History,* ed. L. Hunt (Berkeley: Univ. of California Press, 1989), 1–22. Paula Fass uses the term "sensibilities" throughout her *The Damned and the Beautiful: American Youth in the 1920s* (New York: Oxford Univ. Press, 1977) and Chartier explicitly analyzes the word's utility to historians in Chartier, Georges Duby, Lucien Febvre, Pierre Francastel, & Robert Mandrou, *La Sensibilité dans l'histoire* (Brionne, France: Gérard Monfort, 1987). Simon Frith, "Towards an Aesthetic of Popular Music," in Leppert and McClary, eds., *Music and Society,* 133–49, provides a suggestive model for the sociocultural analysis of popular musics.

7. Dwight Macdonald, "Masscult and Midcult," in *Against the American Grain* (New York: Knopf, 1962), defends high cultural traditions against all

others; while Lawrence W. Levine traces the origins of cultural hierarchies in *Highbrow, Lowbrow: The Emergence of Cultural Hierarchy in America* (Cambridge: Harvard Univ. Press, 1988).

8. J. Ogren, *The Jazz Revolution*, 3–10. F. Scott Fitzgerald, *Tales of the Jazz Age* (New York: Scribner's, 1922). Arnold Shaw, *The Jazz Age: Popular Music in the 1920s* (New York: Oxford Univ. Press, 1987), p. 4, notes a difference between "real jazz" and what Fitzgerald and bandleader Paul Whiteman understood by the term. Houston A. Baker, Jr., *Modernism and the Harlem Renaissance* (Chicago: Univ. of Chicago Press, 1987), 4–6, discusses the narrowness of Fitzgerald's vision of the jazz age.

Chapter One

1. Richard Wang, "Researching the New Orleans–Chicago Jazz Connection: Tools and Methods," *Black Music Research Journal* 8 (1988), 101–2; Leroy Ostransky, *Jazz City: The Impact of Our Cities on the Development of Jazz* (Englewood Cliffs, N.J.: Prentice-Hall, 1978), 59.

2. Dan Morgenstern, "Jazz as an Urban Music," in *Music in American Society, 1776–1976: From Puritan Hymns to Synthesizers,* ed. George McCue (New Brunswick, N.J.: Transaction Books, 1977), 139. Marshall Stearns, *The Story of Jazz* (New York: Oxford Univ. Press, 1956), 112. *Variety* documents the importance of Chicago bands to the national jazz age, e.g.: Dec. 29, 1922, p. 15; Jan. 5, 1923, p. 1; March 8, 1923, p. 16; Aug. 2, 1923, p. 5; June 25, 1924, p. 5; Aug. 6, 1924, p. 37; Sept. 10, 1924, p. 39; also see: Ostransky, *Jazz City,* chap. 6.

3. Indianapolis *Freeman*, March 6, 1915, p. 1; Harold F. Gosnell, *Negro Politicians: The Rise of Negro Politics in Chicago* (Chicago: Phoenix Books, 1967), 74–75, 102–20, 170–71. Dempsey J. Travis, *Autobiography of Black Politics* (Chicago: Urban Research Press, 1987), 35–64.

4. Motts's activities and events at the Pekin Inn are reported in Chicago *Broad Ax,* Feb. 3, 1906, p. 1; March 24, 1906, p. 1; May 12, 1906, p. 2; July 2, 1910, p. 2; Jan. 28, 1911, p. 2; April 15, 1911, p. 1; April 29, 1911, p. 2; May 13, 1911, p. 1; Aug. 10, 1912, p. 2; Sept. 14, 1912, p. 1; Dec. 14, 1912, p. 2. The *Broad Ax,* March 24, 1906, p. 1, reports that the theater will open on March 31st, while the same source on Aug. 10, 1912, p. 2, indicates that it opened on March 17th. Alan H. Spear, *Black Chicago: The Making of a Negro Ghetto, 1890–1920* (Chicago: Univ. of Chicago Press, 1967), 76, identifies Motts as "a community leader." Gosnell, *Negro Politicians,* 127–28, analyzes Motts's political activities. John F. Steiner and Charles A. Sengstock, Jr., "A Chronological Survey of the Chicago *Defender* and the Chicago *Whip* from 1909 to 1930 on the Development of Jazz Music on Chicago's South Side," unpublished typescript, CJA identifies some articles on Motts and the Pekin but the *Broad Ax* is the fullest source. *Variety,*

June 2, 1922, p. 10; New York *Clipper,* July 30, 1924, p. 18; Rudi Blesh and Harriet Janis, *They All Played Ragtime* (New York: Oak Publications, 1971), 154–55.

5. On the sporting fraternity, see Steven A. Reiss, *City Games: The Evolution of American Urban Society and the Rise of Sports* (Urbana: Univ. of Illinois Press, 1989), 14–15. Chicago *Defender,* March 4, 1910, p. 6; March 26, 1910, p. 6; the April 23, 1910, p. 4, issue claims that Motts "caters much at present to white people." The issue of May 7, 1910, p. 3, notes the appearance of a white blackface comedian at the Pekin. Oct. 1, 1910, p. 4; Nov. 26, 1910, p. 3, labels the Pekin as the first "Colored playhouse"; also see *Defender,* May 6, 1911, p. 4; July 29, 1911, p. 4; Aug. 12, 1911, p. 4; Sept. 21, 1912, p. 1; Aug. 28, 1920; June 16, 1923, for further claims concerning the racial orientation of the Pekin Theater. "According to evidence confirmed by U.S. census figures, there is little possibility for a colored business man to make a living solely from the patronage of his own people ... This affords one explanation of the fact that most of his business is of such a character that a white man is willing to patronize it...." Louise de Koven Bowen, *The Colored People of Chicago: An Investigation* (Chicago: Juvenile Protective Association, 1913), 9. Theatrical columnist Sylvester Russell defended the appeal of South Side cabarets to an inter-racial clientele in Indianapolis *Freeman,* May 6, 1916, p. 5, and Aug. 19, 1916, p. 5.

6. Chicago *Broad Ax,* Feb. 3, 1906, p. 1; March 24, 1906, p. 1; April 7, 1906, p. 2; May 12, 1906, p. 2; the July 9, 1910, p. 2, issue identifies Motts as "full of race pride."

7. Chicago *Defender,* March 26, 1910, p. 6; April 23, 1910, p. 4; April 30, 1910, p. 4; May 7, 1910, p. 3; Aug. 6, 1910, p. 4.

8. The history of the Pekin Theater after Motts's death can be traced in the black Chicago newspapers and the Indianapolis *Freeman.* In particular, see *Defender,* July 29, 1911, p. 4; July 26, 1913, p. 6; Oct. 18, 1913, p. 6; Nov. 8, 1913, pp. 1 & 7; Nov. 22, 1913, p. 6; Jan. 31, 1914, p. 7; and Indianapolis *Freeman,* March 6, 1915, p. 5. The "Ghost of Motts" is described in the *Freeman,* Aug. 19, 1916, p. 5.

9. The date of Jackson's first Chicago performances, which is subject to debate, is placed in 1905 by Blesh and Janis, *They All Played Ragtime,* 149–60—which provides no documentation—but is located after 1910 by Lawrence Gushee, "A Preliminary Chronology of the Early Career of Ferd "Jelly Roll" Morton," *American Music* 3 (Winter 1985), 404 fn.38, 411; Gushee insists upon using only the dates at which the Chicago and/or Indianapolis newspapers actually mentioned the presence of a particular musician in Chicago. Such dates can be verified through primary printed sources in a way that earlier dates provided by jazz writers cannot, but they do not rule out, of course, the possibility of an earlier presence. Dempsey J. Travis, *An Autobiography of Black Jazz* (Chicago: Urban Research Institute, 1983), 12.

10. Peyton gives the 1906 date in Chicago *Defender,* Dec. 12, 1925, p. 7. He creates a chronology for the history of Chicago jazz in the following issues of the same paper: Oct. 17, 1925, p. 6; Feb. 5, 1927, p. 6, and the block quote concerning Sweatman appears in the second of these two locations. The *Broad Ax,* July 15, 1911, p. 1, mentions Sweatman's role in Motts's funeral. The New York *Clipper,* June 4, 1919, p. 14, carries Sweatman's advertisement of himself as the originator of ragtime and jazz clarinet playing; the April 19, 1922 issue, p. 22, of the same publication reports on his claim to have made the first jazz records. The *Clipper,* Feb. 28, 1923, p. 24, and *Variety,* Jan. 31, 1920, p. 23, further trace Sweatman's activities.

11. Johnson's political alliance with Wright is mentioned in the Chicago *Defender,* April 9, 1910, p. 3. Jack Johnson and the Cafe de Champion are described in Chicago *Defender,* June 8, 1912, p. 1; June 29, 1912, p. 1; July 6, 1912, p. 6; July 13, 1912, p. 1; April 5, 1913, p. 1; May 17, 1913, p. 1. Randy Roberts, *Papa Jack: Jack Johnson and the Era of White Hopes* (New York: Free Press, 1983), chap. 9, details the factually flimsy and emotionally charged case against Johnson.

12. The role of cabarets in providing an outlet for black entertainers barred by racism from vaudeville theaters is discussed by theatrical columnist Sylvester Russell in the Chicago *Defender,* Aug. 27, 1910, p. 4. Activities at the Pompeii are reported in the Chicago *Defender,* June 15, 1912, p. 2; June 21, 1913, p. 3; May 2, 1914, p. 6; and Aug. 29, 1914, p. 6, which reports on the club's name change and Morton's role as musical director. The Elmwood is also covered in the *Defender,* March 29, 1913, p. 1, and May 2, 1914, p. 6.

13. Jones's career emerges from the pages of Chicago *Broad Ax:* July 15, 1911, p. 1; Dec. 30, 1911, pp. 2, 5; Nov. 1, 1913, p. 1; Nov. 8, 1913, p. 1; July 17, 1915, p. 4; Aug. 21, 1915, p. 4; Sept. 25, 1915, p. 8; Dec. 2, 1916, p. 4; Jan. 20, 1917, p. 1; Feb. 3, 1917, p. 1; Aug. 25, 1917, p. 4; Dec. 7, 1918, p. 2; Dec. 21, 1918, p. 7; May 17, 1919, p. 5; May 22, 1920, p. 3; and the Chicago *Defender,* July 31, 1909, p. 3; June 14, 1913, p. 6; Nov. 8, 1913, pp. 1, 7; Jan. 23, 1915, p. 6; Sept. 4, 1915, p. 6. Jelly Roll Morton and/or Alan Lomax confuse the two Elites, mistaking the second one for the first, Lomax, *Mr. Jelly Roll,* 152.

14. The second Elite Cafe also known as Teenan Jones' Place was heavily promoted in the Indianapolis *Freeman,* Jan. 2, 1915, p. 1; Feb. 6, 1915, p. 5; July 10, 1915, p. 5; July 17, 1915, p. 5; Aug. 21, 1915, p. 5; Nov. 6, 1915, p. 5; Jan. 22, 1916, p. 5; and Feb. 5, 1916, p. 5.

15. Jones's political activities are analyzed in Gosnell, *Negro Politicians,* 128–30, 173.

16. Reports on the DeLuxe Cafe appear in the Indianapolis *Freeman,* Jan. 9, 1915, p. 5; Jan. 23, 1915, p. 8; Feb. 27, 1915, p. 5, announces that there are no fights there; Feb. 19, 1916, p. 5; and March 11, 1916, p. 5. Preer's illness is reported on May 13, 1916, p. 5.

17. Chicago *Defender,* Jan. 23, 1915, p. 6.

18. Gushee, "Early Career of Ferd 'Jelly Roll' Morton," 404. Morton & Lomax, *Mister Jelly Roll: The Fortunes of Jelly Roll Morton, New Orleans Creole and "Inventor of Jazz"* (New York: Duell, Sloan, and Pearce, 1950), 152–55.

19. Chicago *Defender,* May 27, 1916, p. 6; June 17, 1916, p. 4; Sept. 30, 1916, p. 3; Oct. 14, 1916, p. 4. Eileen Southern, *Music of Black Americans: A History,* 2d ed. (New York: W. W. Norton, 1983), 362, identifies the sources above as documentary evidence of the first appearance of the word "jass" in the history of black music; Gunther Schuller, *Early Jazz: Its Roots and Musical Development* (New York: Oxford Univ. Press, 1968), 4. Indianapolis *Freeman,* Oct. 28, 1916, p. 5.

20. William M. Tuttle, Jr., *Race Riot: Chicago in the Red Summer of 1919* (New York: Atheneum, 1970), chap. 3; Allan H. Spear, *Black Chicago,* chap. 8; St. Clair Drake and Horace R. Cayton, *Black Metropolis: A Study of Negro Life in a Northern City* (New York: Harcourt, Brace, 1945), 5–18. James R. Grossman, *Land of Hope: Chicago, Black Southerners, and the Great Migration* (Chicago: Univ. of Chicago Press, 1989), 3–4.

21. John Chilton, *Who's Who of Jazz: Storyville to Swing Street* (Philadelphia: Chilton, 1972), *passim.* For a short but penetrating discussion of the musician as vagabond entertainer, see Jacques Attali, *Noise: The Political Economy of Music,* trans. Brian Massumi (Minneapolis: Univ. of Minnesota Press, 1985), 13–15. Dave Peyton, "The Musical Bunch," Chicago *Defender,* Nov. 19, 1927, p. 10, says that the Creole Band arrived in Chicago in 1911, but he changes the date to 1910 in *Defender,* 6–8–29, p. 10. Frederic Ramsey and Charles Edward Smith echo Peyton's claim for 1911 in "Chicago: 'Every tub its own bottom,'" in *Jazzmen* (New York: Harcourt Brace, 1939), 95–96; Ramsey uses the 1910 date in "Going Down State Street: Lincoln Gardens and Friars Inn Set the Stage for Chicago Jazz," *Jazzways,* ed. George S. Rosenthal and Frank Zachary (New York: Greenberg, 1946), 22; John Steiner accepts a date of 1912 in "Chicago," *Jazz: New Perspectives on the History of Jazz by Twelve of the World's Foremost Jazz Critics and Scholars,* ed. Nat Hentoff and Albert J. McCarthy (New York: Da Capo, 1975), 145. Lawrence Gushee, "How the Creole Band Came To Be," *Black Music Research Journal* 8 (1988), 83–100, disputes these dates and places their arrival in Chicago in Feb. 1915. A 1918 letter from the Creole Band to columnist Tony Langston claims that the band had been "out five years" [on the road] and therefore suggests that they could have played in Chicago as early as 1913, Chicago *Defender,* March 16, 1918, p. 6. On Bechet, see Sidney Bechet, *Treat It Gentle* (New York: Hill & Wang, 1960), chaps. 6–7; and John Chilton, *Sidney Bechet, the Wizard of Jazz* (New York: Oxford Univ. Press, 1987), chap. 4. Frank Gillis and John W. Miner, eds., *Oh, Didn't He Ramble: The Life Story of Lee Collins as Told to Mary Collins* (Urbana: Univ. of Illinois Press, 1974), chaps. 1–2.

22. Attali, *Noise,* 13. Jimmy Wade's Orchestra was among the first to take

South Side jazz from Chicago to New York when, in June 1926, they moved from the Moulin Rouge to New York for a 10-week run at the Club Alabam. Chicago *Defender,* June 12, 1926, p. 6; July 31, 1926, p. 6. A summary of movement to Europe by African-American musicians appears in Dave Peyton, "Ours Musicians in Europe," Chicago *Defender,* Jan. 2, 1926, p. 7.

23. Pay scales are noted in the Chicago *Defender,* July 24, 1926, p. 6; Chilton, *Sidney Bechet,* 27–32; Gene Anderson, "Johnny Dodds in New Orleans," *American Music* 8 (1990), 422; Zutty Singleton Interview, Singleton File, WRHJA; Lil Hardin-Armstrong in *Hear Me Talkin' to Ya: The Story of Jazz as Told by the Men Who Made It,* ed. Nat Shapiro and Nat Hentoff (New York: Dover, 1966), 91.

24. Lil Hardin-Armstrong, "Lil Armstrong Reminisces about Early Chicago Days," *Down Beat* 18 (June 1951), 12, also reprinted in Shapiro and Hentoff, *Hear Me Talkin',* 91. Black immigrants accepted their new, northern world "with gusto"; the Black Belt retained "a general air of optimism" until 1929. Drake and Cayton, *Black Metropolis,* 76, 80. "Amusement Places Reopen in Riot Zones," Chicago *Whip,* Aug. 15, 1919, p. 6. Despite persistent disapproval of the black-and-tan cabarets, no evidence was ever presented that they had played any role in generating the race riot of 1919. Chicago Commission on Race Relations, *The Negro in Chicago: A Study of Race Relations and a Race Riot* (Chicago: Univ. of Chicago Press, 1922), 324. Anonymous immigrant as quoted in Tuttle, *Race Riot,* 79.

25. Tuttle, *Race Riot,* 213–16; Mark Haller, "Urban Vice and Civic Reform: Chicago in the Early Twentieth Century," in *Cities in American History,* ed. Kenneth T. Jackson and Stanley K. Schultz (New York: Knopf, 1972), 299; Grossman, *Land of Hope,* 150. Ostransky, *Jazz City,* 83–91; Spear, *Black Chicago,* 157; Joanne J. Meyerowitz, *Women Adrift: Independent Wage Earners in Chicago, 1880–1930* (Chicago: Univ. of Chicago Press, 1988), 110; Drake and Cayton, *Black Metropolis,* 609.

26. Perry Duis, *The Saloon: Public Drinking in Chicago and Boston, 1880– 1920* (Urbana: Univ. of Illinois Press, 1983), 236; Herman Kogan and Lloyd Wendt, *Chicago: A Pictorial History* (New York: E. P. Dutton, 1958), 174–77.

27. "The Stroll" is described in Donald Spivey, ed., *Union and the Black Musician: The Narrative of William Everett Samuels and the Chicago Local 208* (Lanham, Md.: Univ. Press of America, 1984), 39; Eddie Condon with narration by Thomas Sugrue, *We Called It Music: A Generation of Jazz* (New York: Holt, 1947), 114. Hugues is quoted in Grossman, *Land of Hope,* 117, and also 140, 146. Chicago *Defender,* May 2, 1914, p. 1, also describes the Stroll as does Willie "The Lion" Smith with George Hoefer, *Music on My Mind: Memoirs of an American Pianist* (Garden City, N.Y.: Doubleday, 1964), chap. 12. Chicago *Whip,* Dec. 16, 1922, reports on the installation of electric lights on State Street.

28. Chicago *Defender,* April 9, 1910, p. 1; April 9, 1910, p. 3; and especially

"State Street the Great White Way," May 11, 1912, p. 8; and June 15, 1912, p. 1; May 2, 1914, p. 1, describes the Stroll from a tourist's perspective.

29. (New York: Fred Fisher, 1922), 2, CHS. On the white cultural tradition of voyeuristic fascination with blacks, see Winthrop Jordan, *White Over Black: American Attitudes Toward the Negro, 1550–1812* (Chapel Hill: Published for the Institute of Early American History & Culture by the Univ. of North Carolina Press, 1968). On this same pattern in twenties Chicago, see Commission on Race Relations, *Race in Chicago,* 475, which also states that "Negroes know more of the habits of action and thought of the white group than white people know of similar habits in the Negro group."

30. John F. Kasson, "Organization of Entertainment in the Late 19th and Early 20th Centuries," *Henry Ford Museum & Greenfield Village Herald* 14 (1985), 3–13; Chicago *Defender,* July 21, 1923, p. 6; Lewis A. Erenberg, "'Ain't We Got Fun? Popular Entertainment in Chicago, 1893–1929," *Chicago History* 38 (1985–86), 9.

31. *Black Chicago,* 116. On Joyland Park, see Chicago *Broad Ax,* June 23, 1923, p. 2; June 30, 1923, p. 1; July 7, 1923, p. 2; July 14, 1923, p. 1.

32. Bowen, *The Colored People of Chicago,* 9.

33. Harold B. Segal, *Turn-of-the-Century Cabaret* (New York: Columbia Univ. Press, 1987), 75–76; *Dictionary of American Slang,* comp. and ed. Harold Wentworth and Stuart Berg Flexner (New York: Crowell, 1967), 40–41.

34. Bowen, *The Colored People of Chicago,* 28. Erenberg, *Steppin' Out,* 124–32. Smith, *Music on My Mind,* 127. Travis, *Autobiography of Black Jazz,* 43, states that no South Side Chicago club was closed to blacks but goes on to record that some of them "did have a Jim Crow seating policy."

35. On early Dreamland, see Chicago *Defender,* Sept. 19, 1914, p. 1; Sept. 19, 1914, p. 6; Sept. 26, 1914, p. 5; Oct. 10, 1914, p. 6; June 9, 1917, p. 4; Chicago *Broad Ax,* May 5, 1917, p. 4; Sept. 7, 1918, p. 12; Chicago *Whip,* June 24, 1919, p. 1; July 19, 1919, p. 5; Aug. 15, 1919, p. 6; Oct. 4, 1919, p. 2; and Travis, *Autobiography of Black Jazz,* 28. On Oliver, see Frederic Ramsey, Jr., "King Oliver," in *Jazzmen,* chap. 3; Walter C. Allen and Brian A. L. Rust, *King Joe Oliver* (London: Sidgwick & Jackson, n.d.)—a revised edition of which has been published as Laurie Wright, *Walter C. Allen & Brian A. L. Rust's "King" Joe Oliver* (London: Storyville Publications, 1987); Martin T. Williams, "Papa Joe," in *Jazz Masters of New Orleans* (New York: DaCapo, 1979), 79–120; Edmond Souchon, "King Oliver," in *Jazz Panorama: From the Pages of the Jazz Review* (New York: Crowell-Collier, 1962). On Armstrong and Lil Hardin-Armstrong's Dreamland Syncopators, see Chicago *Defender,* Oct. 31, 1925, p. 6; Nov. 7, 1925, p. 8; and James Lincoln Collier, *Louis Armstrong: An American Genius* (New York: Oxford Univ. Press, 1983), 157. For added geographical clarity about the location of Chicago's jazz clubs, readers can consult the map called "Chicago Jazz Spots, 1914–1928 redrawn from the Original Map by Paul Eduard Miller

and Richard M. Jones," in *Esquire's 1946 Jazz Book*, ed. P. E. Miller (New York: Smith & Durrell, 1946).

36. "New Ideas," Chicago *Defender*, Feb. 2, 1918, p. 4; "Dreamland Most Beautiful Cafe Owned and Controlled by Race Interests," Chicago *Defender*, Oct. 10, 1925, p. 6.

37. Chicago *Defender*, Oct. 18, 1924, p. 7. South Side resentment of white black-and-tan entrepreneurs in their neighborhood is echoed in Chicago *Broad Ax*, May 26, 1923, p. 1. Eric Lott, "'The Seeming Counterfeit': Racial Politics and Early Blackface Minstrelsy," *American Quarterly* 43 (1991), 225, introduces the insightful expression "commercial regulation of black cultural practices."

38. Chicago *Defender*, Nov. 3, 1923, p. 8; June 21, 1924, p. 7; July 5, 1924, p. 6; July 12, 1924, p. 6; Milt Hinton Interview with David Berger, ts., IJS. On the Abbott-Bottoms partnership, see Roi Ottley, *The Lonely Warrior: The Life and Times of Robert S. Abbott* (Chicago: Regnery, 1955), 201.

39. On Virgil Williams and the Royal Gardens, see Chicago *Broad Ax*, June 5, 1920, p. 1; Chicago *Whip*, Dec. 23, 1922, p. 1; Williams's references to Florence Majors are confirmed by Interview with Stella [Mrs. Joseph] Oliver, ms., Hogan Jazz Archive. However, the same person is called "Isabelle" Majors in Al Monroe, ts. notes, Chicago Joints File, JSC.

40. Baby Dodds as told to Larry Gara, *The Baby Dodds Story* (Los Angeles: Contemporary Press, 1959), 35. New York *Clipper*, Sept. 14, 1923, p. 31. JPA support of the Lincoln Gardens is announced in *Annual Report of the Juvenile Protective Association* (Chicago: JPA, 1924), 12–13. Martin T. Williams, *Jazz Masters of New Orleans*, 106, asserts that gangsters owned the Lincoln Gardens. Bud Freeman comments on his reception there in Don DeMichael and Wayne Jones, Institute of Chicago Jazz Oral History Interview with Bud Freeman, Nov. 3, 1980, audio tape, CJA. Wettling's description of the dance hall appears in *Hear Me Talkin' to Ya*, 99–100. Condon, *We Called It Music*, 94.

41. Chicago *Defender*, May 10, 1919, p. 9; June 28, 1919, p. 8; July 2, 1927, p. 8; July 23, 1927, p. 6; Oct. 8, 1927, p. 8; Jan. 21, 1928, p. 9. Frederic Ramsey, Jr., "King Oliver and His Creole Jazz Band," in *Jazzmen*, 76.

42. Glaser, described as "a Chicago dealer in used cars," is reported to have owned the Green Mill Gardens. *Variety*, Feb. 15, 1923, p. 21. The Sunset was attacked by the state attorney general as "a common, ill-governed, disorderly house given to the encouragement of idleness, drinking, and licentious and lascivious conditions," but nothing came of the charge. *Variety*, May 12, 1922, p. 19; Nov. 17, 1922, p. 38; Collier, *Louis Armstrong*, 160–62, 166, 176, 272, 344, 350, describes Glaser and his relations with Armstrong. The Hinton interview with Berger (IJS) discusses the bassist's contacts with Glaser. On musical entertainment at the Sunset Cafe, see Chicago *Defender*, Sept. 10, 1921, p. 6; March 31, 1923, p. 6; July 7, 1923, p. 6; Dec. 26, 1925, p. 6; April 17, 1926, p. 6; Aug. 28, 1926, p. 6; Jan. 8, 1927, p. 6; May 7, 1927, p. 8; July 2, 1927, p. 8; July 23, 1927,

p. 6; Sept. 17, 1927, p. 9. Stanley Dance, *The World of Earl Hines* (New York: Scribner's, 1977), 45–46; "Wild Bill" Davison Interview with Hal Willard, ts., IJS.

43. *Variety,* May 24, 1923, p. 10, and Chicago *Broad Ax,* May 26, 1923, p. 1, announce the closing of Al Tierney's, while *Variety,* Dec. 2, 1925, p. 47, and April 21, 1926, p. 45, describe the Plantation Club scene in some detail. *Jazz Grove* II, 202, identifies the owners. *People of the State of Illinois (ex rel. Irving Cohen) v. City of Chicago,* Cook County Superior Court transcript, Dec. 6, 1924, CHS. Someone has penciled "Sunset Amusement Corp." on this document, even though no such organization is mentioned in the text. This would seem to suggest that the Sunset Cafe and the Plantation Cafe were owned by the same company.

44. Travis, *Autobiography of Black Jazz,* 33; manuscript notes, Jimmie Noone File, JSC; *Jazz Grove* II, 195; Interview with Tut Soper by Wayne Jones and Warren Plath, July 19, 1983, ts., CJA. Chicago *Defender,* July 2, 1927, p. 8; Sept. 17, 1927, p. 9. Wesley, M. Neff, "Jimmie Noone," *Jazz Information* (Oct. 4, 1940), 6–9.

45. "Colored Citizens Alarmed Over Social Evil Coming into Their Residence District," Indianapolis *Freeman,* March 11, 1916, p. 1; "Mayor's Actions in Dealing with Vice Seem To Be Wrought with Astounding Inconsistency," Indianapolis *Freeman,* April 1, 1916, p. 1.

46. The Entertainers Cafe was closed in Feb. 1921 by the U.S. Circuit Court of Appeals for violation of the Volstead Act; it reopened one year later. The dancing of its entertainer Julie Rector was declared to be obscene at the same time. *Variety,* April 7, 1922, p. 18; March 29, 1923, p. 47; JPA Files: #92—"Commercialized Prostitution, Personal File of Jesse Binford [1923]"; #93—"Commercialized Prostitution in Chicago, 1922"; #94—"Commercialized Prostitution in Chicago, 1922–24"; #96—"Commercialized Prostitution, 1923–24"; #108—Vice Investigations, 1920–23, specifically indicates that the April 1923 closing of the bordellos had created an "Abnormal Saturday night situation"; #110—Vice Investigations Chicago, 1922–23, UICSC. Dance, *World of Earl Hines,* 45–46.

47. The emphasis to the description of the appeal of cabaret entertainment has been added. See JPA File #92. Sylvester Russell, who covered the South Side entertainment scene for the Indianapolis *Freeman* before the war, had similarly insisted that "proprietors of these places [the two Elites, the Panama, the De-Luxe, and others] are not catering for white prostitutes nor traffic traders. Good order and recreation is their chief aim in catering for public business." *Freeman,* May 6, 1916, p. 5.

48. T.J.Jackson Lears, *No Place of Grace: Antimodernism and the Transformation of American Culture, 1880–1920* (New York: Pantheon, 1981), 52.

49. John F. Szwed, "Race and the Embodiment of Culture," *Ethnicity* 2

(1975), 19–33. Attali, *Noise,* 20–23. Victor Turner, *From Ritual to Theatre: The Human Seriousness of Play* (New York: PAJ Publications, 1982), 20–60.

50. New York *Clipper,* July 21, 1920, p. 6; Aug. 4, 1920, p. 1; March 2, 1921, p. 32; April 6, 1921, p. 30; April 13, 1921, p. 7.

51. Chicago *Defender,* Sept. 4, 1926, p. 6; May 19, 1923, p. 7; Dec. 8, 1923, p. 1; Oct. 10, 1925, p. 6; Sept. 11, 1926; on the NAACP at Dreamland, see *Defender,* July 3, 1926, p. 6.

52. Chicago *Broad Ax,* July 20, 1912, p. 1; Aug. 10, 1912, p. 1; March 10, 1917, p. 4; March 17, 1917, p. 1; Nov. 16, 1918, p. 8; Nov. 25, 1922, p. 2; Sept. 22, 1923, p. 2; Dec. 2, 1922, p. 3; May 26, 1923, p. 1.

53. "Patronize Worthy Race Enterprises Along 'the Stroll,'" Chicago *Defender,* March 20, 1915, p. 9; May 8, 1915, p. 4; May 29, 1915, p. 8; Aug. 7, 1915, p. 8; Sept. 4, 1915, p. 6. Ottley, *The Lonely Warrior,* 100–101.

54. Chicago *Defender,* Aug. 23, 1924, p. 4.

55. *Ibid.,* June 12, 1926, p. 6.

56. Chicago *Broad Ax,* July 9, 1910, p. 2; July 9, 1921, pp. 1–2; July 16, 1921, p. 1; May 6, 1922, p. 1.

57. *Ibid.,* April 21, 1917, p. 4; April 28, 1917, p. 4.

58. *Ibid.,* Dec. 25, 1920, p. 1; Chicago *Defender,* Feb. 21, 1920, p. 12.

59. Chicago *Broad Ax,* June 16, 1923, p. 1; June 23, 1923, p. 2; June 30, 1923, p. 1; July 7, 1923, p. 2; July 14, 1923, p. 1.

60. As quoted in Lloyd Wendt and Herman Kogan, *Big Bill of Chicago* (New York: Bobbs-Merrill, 1953), 147–69, 168–233. The Chicago *Broad Ax,* Dec. 20, 1919, p. 3, claimed that Harding owned 3,000 houses, flats, and stores on the South Side.

61. Chicago *Broad Ax,* Nov. 29, 1919, p. 1; Dec. 13, 1919, p. 1; Dec. 20, 1919, pp. 3, 8; June 5, 1920, p. 1; June 12, 1920, p. 2.

62. Thompson's mayoral power over cabaret licenses is documented in *Proceedings of City Council, 1917–18* (Chicago: City of Chicago, 1918), 2510–11. Herbert Asbury, *The Great Illusion: An Informal History of Prohibition* (Garden City, N.Y.: Doubleday, 1950), chap. 14. On Dreamland's difficulties in 1923–24, see Chicago *Defender,* Nov. 3, 1923, p. 8; Dec. 8, 1923, p. 1; Dec. 15, 1923, p. 3; June 21, 1924, p. 7; July 5, 1924, p. 6.

63. Chicago *Defender,* Aug. 18, 1917, p. 4; *Variety,* July 14, 1922, p. 11; Dance, *World of Earl Hines,* 52.

64. *Variety,* Jan. 20, 1922, p. 9; March 24, 1922, p. 19; Oct. 13, 1922, p. 8. Erenberg, *Steppin' Out,* 235–36, analyzes how urban prohibition served to enhance the aura of night-club entrepreneurs like Texas Guinan.

65. The closing of Al Tierney's Auto Inn is discussed in Chicago *Defender,* May 26, 1923, p. 2. The struggles of Izzy Shorr and Joe Gorman of the Entertainers Cafe with reformers are traced in the *Defender,* March 19, 1921; Oct. 11, 1924, p. 6; May 23, 1925, p. 7; Dec. 18, 1926, p. 1. Dreamland's adaptation to

reform pressures can be traced in *Defender*, Nov. 3, 1923, p. 8; Dec. 8, 1923, p. 1; Dec. 15, 1923, p. 3; Dec. 29, 1923, p. 6; June 21, 1924, p. 7; July 5, 1924.

66. John R. Schmidt, "Dever of Chicago: A Political Biography," Ph.D. diss., Univ. of Chicago, 1983, pp. 125–26, 164, 176, 246–50; Douglas Bukowski, "William Dever and Prohibition: The Mayoral Elections of 1923 and 1927," *Chicago History* 7 (1978), 109–18; Asbury, *Great Illusion,* 301; Walter C. Reckless, *Vice in Chicago* (Chicago: Univ. of Chicago Press, 1933), 113, offers statistics on the number of places closed and licensed.

67. Chicago *Defender*, June 18, 1927, p. 8.

68. John Steiner, "Chicago," in *Jazz: New Perspectives on the History,* ed. Nat Hentoff and Albert J. McCarthy (New York: Da Capo, 1975), 145; *Jazz Grove* II, 196; and Wang, "New Orleans-Chicago Jazz Connection," 106–8, all date the Elgar-associated group at the Arsonia in 1915. On Elgar's activities, also see: Hennessey, "Jazz Age to Swing," 101–2; Charles Elgar Interview, HJATU. "Negro Music and Musicians," HCWB. The early activities of black musicians in white Chicago are also traced in Paul Eduard Miller, "Thirty Years of Chicago Jazz," *Esquire's 1946 Jazz Book,* ed. P. E. Miller (New York: Smith & Durell, 1946), 8. Fritzel and the Arsonia Cafe are described in Louis M. Starr, "45 Years of Night Life," Chicago *Sun,* June 6, 1943, p. 47, and Fritzel obituary, Chicago *Daily Tribune,* Sept. 29, 1956, pt. 3, p. 12. Chicago *Herald,* Feb. 3, 1916, p. 16.

69. The Dreamland Ballroom is described in Irle Waller, *Chicago Uncensored: Firsthand Stories About the Al Capone Era* (New York: Exposition Press, 1965), 61. Doc Cook's band can be traced in Chicago *Defender,* Oct. 10, 1925, p. 7; Oct. 17, 1925, p. 6; Feb. 18, 1926; March 6, 1926, p. 6; May 22, 1926, p. 6; June 12, 19, 1926, p. 6; March 12, 1927, p. 10; April 9, 1927, p. 9; May 7, 1927, p. 8.

70. Chicago *Herald,* Jan. 31, 1916, p. 16; this club is listed as "Del' Abe Cafe (De Labbie)" in *Jazz Grove,* II, 197.

71. Spivey, ed., *Union and the Black Musician.* On the Moulin Rouge, see *Variety,* Jan. 20, 1922, p. 9; Jan. 27, 1922, p. 9; March 24, 1922, p. 19. Efforts of the white musicians' union to limit black musicians to South Side jobs are detailed by Dave Peyton in the Chicago *Defender,* Oct. 31, 1925, p. 6; Jan. 9, 1926, p. 6; Sept. 18, 1926, p. 6; March 21, 1927, p. 8; April 2, 1927, p. 8; Oct. 8, 1927, p. 8.

72. Ramsey, "Going Down State Street," 32. Stanley Dance Interview with Marge and Zutty Singleton, May 1975, ts. IJS. Advertisements for Kelly's Stables trace its entertainment policy and the sequence of groups which played there. See *This Week in Chicago* 5 (Nov. 19, 1922), 4; 9 (Nov. 9, 1924), 20; 10 (March 1, 1925), 20; 10 (March 8, 1925), 20, CHS; also Theater Programs, 1919–27, CJA. Advertisements and brief descriptions of the club also appear in *Variety,* 1920–28 *passim;* collector and jazz expert John Steiner discusses Bert Kelly as the possible originator of the word "jazz" in Chicago. See "Notes B," *Jazz Odyssey II: The Sound of Chicago, 1923–1940,* Columbia Record CL3L32.

73. Muggsy Spanier Interview, ms., WRHJA. Charles Edward Smith, "White New Orleans," in Ramsey and Smith, eds., *Jazzmen*, 56–57; Chicago *Defender*, Sept. 18, 1926, p. 6.

74. Chicago *Defender*, March 13, 1926, p. 6; Sept. 18, 1926, p. 6.

75. Chicago *Defender*, Jan. 30, 1926, p. 6; Feb. 13, 1926, p. 6; Feb. 27, 1926, p. 6; July 30, 1927, p. 8; Aug. 6, 1927, p. 8. Dominic Pacyga and Ellen Skerrett, *Chicago: City of Neighborhoods* (Chicago: Loyola Univ. Press, 1986), 9–10. *Variety*, March. 17, 1922, p. 19.

Chapter Two

1. Gunther Schuller, *Early Jazz: Its Roots and Musical Development* (New York: Oxford Univ. Press, 1968), 86. For a brief consideration of changes in Oliver's New Orleans style during his Chicago years, also see Thomas Hennessey, "From Jazz Age to Swing: Black Musicians and Their Music, 1917–1935," PH.D. diss., Northwestern Univ., 1973, p. 63. Chicago jazz's new institutional environment is noted in John Chilton, *Sidney Bechet, the Wizard of Jazz* (New York: Oxford Univ. Press, 1987), 31.

2. Kathy J. Ogren, *The Jazz Revolution: Twenties America & the Meaning of Jazz* (New York: Oxford Univ. Press, 1989), 40. Lawrence W. Levine, *Black Culture and Black Consciousness: Afro-American Folk Thought from Slavery to Freedom* (New York: Oxford Univ. Press, 1977), 238.

3. Danny Barker, *A Life in Jazz*, ed. Alyn Shipton (New York: Oxford Univ. Press, 1986), 45. Edmond Souchon, "King Oliver: A Very Personal Memoir," in *Jazz Panorama: From the Pages of the Jazz Review*, ed. Martin T. Williams (New York: Crowell-Collier, 1962), 21–30.

4. The invaluable interviews with South Side musicians conducted by the late John Lax included questions about the status of jazz musicians within the South Side community. John Lax Interview with Willie Randall, Dec. 28, 1971, ts., COHC; Lax Interview with Red Saunders, Dec. 24, 1971, ts., COHC; Lax Interview with Scoville Brown, Dec. 14, 1971, ts., COHC. Jeff Titon, *Early Downhome Blues* (Urbana: Univ. of Illinois Press, 1977), chap. 7, discusses stereotypes of urban blacks. Frederic Ramsey, Jr., describes the small town rustication of arriving New Orleans musicians in "Going Down State Street: Lincoln Gardens and Friars Inn Set the Stage for Chicago Jazz," in *Jazzways*, ed. George S. Rosenthal and Frank Zachary (New York: Greenberg, 1946), 22, 28. Barker, *A Life in Jazz*, 62, describes Chris Kelly as does Lee Collins in *Oh, Didn't He Ramble: The Life of Lee Collins as Told to Mary Collins*, ed. Frank J. Gillis and John W. Miner (Urbana: Univ. of Illinois Press, 1974), 55; the *Broad Ax*, March 10, 1917, p. 4; March 17, 1917, p. 1; and Dec. 2, 1922, p. 3, calls for a greater effort at urbanization and personal hygiene.

5. Ramsey describes Bacquet's theatrical style of dress in "Going Down State

Street," 22, and the Chicago *Defender* June 18, 1910, p. 3, reported on "a success-ful vaudeville actor" who appeared at midnight on the Stroll wearing "his famous diamond horseshoe." Derrick Stewart-Baxter, *Ma Rainey and the Classic Blues Singers* (New York: Stein and Day, 1970), 16, discusses the diamond tooth implants.

6. The cabaret and the tuxedo are discussed in "Cabarets of Now," *Variety,* Dec. 29, 1922, p. 15. Photographs of Chicago jazz musicians of the 1920s appear in Frank Driggs and Harris Lewine, eds., *Black Beauty, White Heat: A Pictorial History of Classic Jazz* (New York: William Morrow, 1982), 49–90. The following sources describe some dimensions of dress among South Side musicians: Ed-mond Souchon, "King Oliver," 21–30; James Lincoln Collier, *Louis Armstrong: An American Genius* (New York: Oxford Univ. Press, 1983), 124–25; Dave Peyton, "The Musical Bunch," Chicago *Defender,* Nov. 3, 1917, p. 12; Nov. 19, 1927, p. 10; Albertson, "Louis Armstrong," booklet packaged with *Giants of Jazz: Louis Armstrong* (Alexandria, Va.: Time-Life Records, 1978), 12–17. Stanley Dance, *The World of Earl Hines* (New York: Scribner's, 1977), 22.

7. Chicago *Defender,* Sept. 24, 1927, pp. 6, 8; Jan. 21, 1928, p. 8.

8. Lawrence Gushee, "How the Creole Band Came to Be," *Black Music Research Journal* 8 (1988), 83–100. Chicago *Defender,* Aug. 7, 1915, p. 8; Feb. 25, 1922, pt. 2, p. 3.

9. "The Musical Bunch," Chicago *Defender,* March 20, 1926, p. 6; March 5, 1927, p. 6. Dance, *World of Earl Hines,* 62, 85. Ramsey, "Going Down State Street," 22. Louis Armstrong, *Satchmo: My Life in New Orleans* (New York: Prentice-Hall, 1954), 194–200; Collins, *Oh, Didn't He Ramble,* 57–58; Baby Dodds as told to Larry Gara, *The Baby Dodds Story* (Los Angeles: Contemporary Press, 1959), 25–28.

10. "Junius C. Cobb," ts. career summary, Junie Cobb File, JSC. Manuscript notes, Jimmie Noone File, JSC. Interview with Tut Soper by Wayne Jones and Warren Plath, July 19, 1983, ts., CJA.

11. Souchon in *Jazz Panorama;* Chicago *Defender,* Nov. 2, 1918, p. 6.

12. Wellman Braud as quoted in Frederic Ramsey, Jr., *Chicago Documen-tary: Portrait of a Jazz Era* (London; Jazz Sociological Society, 1944), 7; Ramsey, "Joe Oliver," in *Jazzmen,* 66.

13. As quoted in Ramsey, "King Oliver," 68.

14. For a helpful analysis of the movement from oral folk thought to liter-acy, see Walter J. Ong, *Orality and Literacy: The Technologizing of the Word* (New York: Methuen, 1982), chap. 2.

15. W. C. Handy, *Father of the Blues,* ed. Arna Bontemps (New York: Macmillan, 1941), 76–77; Dave Peyton, "Career of W. C. Handy," Chicago *Defender,* April 17, 1926, p. 6. Hennessey, "From Jazz Age to Swing," 19, documents James Europe's primitivist deceptions. Collins, *Oh, Didn't He Ramble,* 68–69.

16. Chicago *Defender,* Aug. 15, 1923, p. 7.

17. *Variety,* Jan. 5, 1923, p. 35; July 12, 1923, p. 5. *Talking Machine World,* Dec. 15, 1924, p. 179. Driggs and Lewine, eds., *Black Beauty, White Heat,* 49–90. The photo of Cook's Orchestra appeared in Chicago *Defender,* April 12, 1924, p. 11. Louis Armstrong, *Satchmo,* 213.

18. Wellman Braud, as quoted in Ramsey, *Chicago Documentary,* 7–8.

19. *Variety,* June 2, 1922, p. 17; June 16, 1922, p. 23; June 23, 1922, p. 30; Nov. 10, 1922, p. 7. Collins, *Oh, Didn't He Ramble,* 52–53; Barker, *A Life in Jazz,* 44–45.

20. Dance, *World of Earl Hines,* 30–56.

21. Handbill as reprinted in Walter C. Allen and Brian A. L. Rust, *King Joe Oliver* (London: Sidgwick & Jackson, n.d.), 19.

22. The quote appears in "Whites Framing Colored Shows for Colored Folks," *Variety,* April 16, 1924, p. 47, but also see: *Variety,* March 24, 1922, p. 12; "Cabarets of Now," *Variety,* Dec. 29, 1922, p. 15; April 16, 1924, p. 47.

23. Chicago *Defender,* March 3, 1923, p. 6.

24. Interview with Johnny St. Cyr, ms., Los Angeles, Aug. 27, 1958, WRHJA; Dance, *World of Earl Hines,* 62.

25. Morton as quoted in Ramsey, "King Oliver and His Creole Jazz Band," 71.

26. John Wilson, "Notes on the Music," *Louis Armstrong* (Alexandria, Va.: Time-Life Records, 1978), 33.

27. Hardin as quoted in *Hear Me Talkin' to Ya: The Story of Jazz as Told by the Men Who Made It,* ed. Nat Shapiro and Nat Hentoff (New York: Dover, 1966), 93. Sally Placksin, *American Women in Jazz, 1900 to the Present: Their Words, Lives, and Music* (New York: Wideview Books, 1982), 59. Gene Anderson claims that clarinetist Johnny Dodds "possessed a superb musical memory." "Johnny Dodds in New Orleans," *American Music* 8 (1990), 415.

28. *Variety,* Oct. 1, 1924, p. 28. Pianist Art Hodes describes his accompaniment of vocalists as a musical learning experience in "The Rainbow Cafe," *Selections from the Gutter: Portraits from the "Jazz Record,"* ed., Hodes and Chadwick Hansen (Berkeley: Univ. of California Press, 1977), 10.

29. Peyton's ad appears in the Chicago *Defender,* Nov. 29, 1913, p. 6. Walter Ong argues that when literacy first intrudes into oral cultures, writing becomes a craft, "a trade practiced by craftsmen whom others hire to write a letter or a document." *Orality and Literacy,* 94.

30. Commission on Chicago Historical and Architectural Landmarks, "Black Metropolis Historical District" (Chicago: CCHAL, 1984), 2–7.

31. David Evans, *Big Road Blues: Tradition and Creativity in the Folk Blues* (Berkeley: Univ. of California Press, 1982), 60–64. The writing and publishing of orally transmitted music is described in Stephen Calt, "Paramount, Part 2: The Mayo Williams Era," *78 Quarterly* 1 (1989), 10–30. The question of musical literacy should be seen as one specialized dimension of the more general question

of the impact of literacy on African-American cultural sensibilities. Again, see Lawrence Levine, *Black Culture and Black Consciousness,* 155–58, 177.

32. Chicago *Defender* Jan. 8, 1916, p. 7; Feb. 22, 1919, p. 13; Feb. 21, 1920, p. 7; March 20, 1920, p. 7; and Williams is quoted in Chicago *Whip* Sept. 27, 1919, p. 11.

33. Ramsey, "King Oliver," 96, quotes from Oliver's letter to Petit.

34. On Morton, see Jelly Roll Morton and Alan Lomax, *Mr. Jelly Roll: The Fortunes of Jelly Roll Morton, New Orleans Creole and "Inventor of Jazz"* (New York: Duell, Sloan, & Pearce, 1950), 184–88; and Wright, *Allen & Rust's "King" Oliver,* 337. The information was gathered from *Talking Machine World* and contributed to this revised edition by Lawrence Gushee.

35. Dave Peyton, "The Musical Bunch—Songwriting Business," Chicago *Defender,* Jan. 23, 1926, p. 6.

36. Tate's minstrel version of "Royal Garden Blues" is described in the Chicago *Defender,* Feb. 20, 1926, p. 6.

37. *Ibid.,* May 17, 1919, p. 9; July 27, 1918, p. 7.

38. *Ibid.,* March 31, 1923, p. 6 describes the racetrack theme at the Sunset Cafe; also see May 24, 1919, p. 8; March 18, 1922, p. 6; Chicago *Whip,* July 19, 1918, p. 5.

39. Milt Hinton describes Bill Johnson's and his own musical clowning in an Interview with Tom Piazza, Washington, D.C., Jan. 1977, ts., IJS. Lawrence Duhe as quoted in Chilton, *Sidney Bechet,* 28. Chicago *Defender,* Aug. 30, 1924, p. 6, reports on Dunn; Jimmy Bertrand's Washboard Wizards, Okeh Records, Chicago April 25, 1929; Dempsey J. Travis, *An Autobiography of Black Jazz* (Chicago: Urban Research Institute, 1983), 34.

40. Louis Armstrong, "Life Story of Louis Armstrong," thermofax ts., IJS.

41. The quote about musical instruments and technology comes from Ajay Heble, "The Poetics of Jazz from Symbolists to Semiotics," *Textual Practice* 2 (1988), 51–68. The issue also arises in Theodor W. Adorno, "Perennial Fashion—Jazz" (review of Wilder Hobson, *American Jazz Music* and Winthrop Sargeant, *Jazz: Hot and Hybrid*), in Adorno, *Prisms,* trans. Samuel and Shierry Weber (Cambridge, Mass.: MIT Press, 1981), 121–32, which hides interesting points concerning jazz and technology in what otherwise amounts to an undocumented diatribe against "jazz"; Adorno often elides "jazz" and dance music.

42. Ong, *Orality and Literacy,* 83.

43. Charles Cooke's doctorate is discussed in *Defender,* June 19, 1926, p. 6. An interview/biographical sketch of Clifford King appears in the Illinois Writers Project of the W.P.A., "Negro Music in Chicago," HCWB; the Chicago *Defender,* Dec. 4, 1926, reports on the musical education of Jerome Pasquall; on Simeon, see "Omer Simeon," *Jazz Information* II (July 26, 1940), 7–10.

44. N. Clark Smith's Chicago activities are described and photographed in Chicago *Broad Ax,* Sept. 1, 1906, pp. 1, 2; Dec. 29, 1906, pp. 3, 5, 7; June 28, 1913,

p. 3; July 5, 1913, p. 2; Chicago *Defender* July 9, 1921, p. 4; Nov. 12, 1921, p. 3; Oct. 24, 1925, p. 6; Nov. 28, 1925, p. 6; Dec. 19, 1925, p. 6; interview with Milt Hinton in Dance, *The World of Earl Hines;* Tom Piazza Interview with Hinton, ts., Wash., D.C., 1977, 8, 36–38, 44, IJS.

45. Interview with Albert "Happy" Caldwell, ts., IJS; Ford S. Black, *Black's Blue Book, 1917. Directory of Chicago's Active Colored People and Guide to Their Activities* (Chicago: Ford S. Black, [1917]), 41–43; Ford S. Black, *Black's Blue Book, Business and Professional Directory* ([Chicago]: Ford S. Black, 1921). Erskine Tate vertical file, IJS; Interview with Jimmy Bertrand, Sept. 9, 1959, ts., WRHJA.

46. Louis Armstrong, *Swing That Music* (New York: Longmans Green, 1936), 71; Lil Hardin as quoted in Ramsey, *Chicago Documentary,* 6–7. East Coast pianist Willie "The Lion" Smith played on Chicago's South Side for a time in 1923; he emphasized in his memoirs that wind instrumentalists played a more prominent role there than in New York, where the pianists starred. George Hoefer, *Music on My Mind: The Memoirs of an American Pianist* (Garden City, N.Y.: Doubleday, 1964), chap. 12. John Steiner, "Chicago," in *Jazz: New Perspectives on the History of Jazz* (New York: Da Capo, 1975), 155–56, lists Chicago pianists.

47. Don DeMichael, "Percussion's Dean Steps to a Lively Beat," Chicago *Tribune* (Nov. 12, 1978), 23, 26. Interview with Caldwell, IJS. Interview with Bertrand, WRHJA.

48. Chadwick Hansen, "Social Influences on Jazz Style: Chicago, 1920–1930," *American Quarterly* 12 (1960), 493. This important article first conceptualized the problem of Chicago jazz's interactions with historical and cultural forces and demonstrated the richness of the Chicago *Defender* as a primary printed source.

49. The quotation comes from Peter Burke, *Popular Culture in Early Modern Europe* (New York: New York Univ. Press, 1978), 60. For a concise summary of the problems with traditional distinctions between "producers" and "consumers" of culture, and the ways in which "popular culture" adapts fine arts culture to its uses, see Roger Chartier, *Cultural History: Between Practices and Representations,* trans. Lydia G. Cochrane (Ithaca: Cornell Univ. Press, 1988), 39–41; Richard Hoggart, *The Uses of Literacy: Aspects of Working Class Life with Special Reference to Publications and Entertainment* (London: Chatto and Windus, 1957), 27–29, 142.

50. Darnell Howard's musical education can be reconstructed through Chicago *Defender,* Dec. 16, 1911, p. 7; May 25, 1912, p. 5; Chicago *Broad Ax,* May 18, 1912, p. 1; and Albert J. McCarthy, "Darnell Howard—Pt. 1," *Jazz Monthly* 6 (1960), 7–9. Elgar's and Howard's activities are traced in the Chicago *Broad Ax,* May 18, 1912, p. 1; May 18, 1918, p. 1; May 25, 1918, p. 1; Dec. 31, 1921, p. 1; Jan. 1, 1921, p. 1; and McCarthy, "Darnell Howard—Pt. 1," 7–9.

51. Thomas Hennessey, "From Jazz Age to Swing: Black Musicians and Their Music, 1917–1935," Ph.D. diss., Northwestern Univ., 1973; Thomas J. Hennessey, "The Black Chicago Establishment 1919–1930," *Journal of Jazz Studies* 2 (Dec. 1974), 37. Hsio Wen Shih, "The Spread of Jazz and the Big Bands," in *Jazz: New Perspectives on the History of Jazz,* ed. Nat Hentoff and Albert J. McCarthy (New York: Da Capo, 1975), 173–87.

52. Garvin Bushell as told to Mark Tucker, *Jazz from the Beginning* (Ann Arbor: Univ. of Michigan Press, 1988), 25–26.

53. None of them explains exactly what "spelling" meant in this context, but they had likely learned to name the lines and spaces of the grand staff with their appropriate letters. According to Mary Lou Williams, the Kansas City pianist/arranger, those musicians who could only "spell" were "slower than the good readers," which seems to indicate that they also could play the notes on their instruments (at slow tempi); as quoted in Nathan W. Pearson, Jr., *Goin' to Kansas City* (Urbana: Univ. of Illinois Press, 1987), 62. Ong, *Orality and Literacy,* 100.

54. "Bad Habits," Chicago *Defender,* Jan. 30, 1926, p. 6; "Song Writing Business," *ibid.,* Jan. 23, 1926, p. 6; "The Orchestra Conductor," *ibid.,* Feb. 13, 1926, p. 6; "Standard Music," *ibid.,* June 5, 1926, p. 6; "Things in General," *ibid.,* Aug. 7, 1926, p. 6; "Interested Music Teachers," *ibid.,* July 17, 1926, p. 6; "Thoughtful Musicians," *ibid.,* July 24, 1926, p. 6; "Characteristics in the Orchestra," *ibid,* July 3, 1926, p. 6; "Our Spirituals," *ibid.,* March 6, 1926, p. 6; "The Orchestra," *ibid.,* April 3, 1926, p. 6; "Things in General," *ibid.,* Jan. 29, 1927, p. 6; "Things in General," *ibid.,* March 5, 1927, p. 6; March 12, 1927, p. 10.

55. The cigarette incident is reported by Peyton in the Chicago *Defender,* April 16, 1927, p. 8, and repeated in Ramsey, "King Oliver," 76–77. Walter C. Allen and Brian A. L. Rust, *King Joe Oliver* (London: Sidgwick and Jackson, n.d.), 15–16, interpret the confrontation as part of a musicians' duel over an important job, a view confirmed more recently by Barney Bigard, *With Louis and the Duke,* ed. Barry Martyn (New York: Oxford Univ. Press, 1986), 28–29. Peyton's Symphonic Syncopators (two cornets, three saxes, trombone, banjo, tuba, drums, and piano) were reviewed in *Variety,* Nov. 12, 1924, p. 31.

56. Allen and Rust, *King Joe Oliver,* 15–22; Ramsey and Smith, *Jazzmen,* 76; Williams, *Jazz Masters of New Orleans,* 100–101; Collins, *Oh, Didn't He Ramble,* 66; Chicago *Defender,* Jan. 29, 1927, p. 6.

57. Chicago *Defender,* Jan. 23, 1926, p. 6; also June 19, 1926, p. 6; March 19, 1927, p. 8.

58. Ralph Ellison, "Living with Music," *Shadow and Act* (New York: Random House, 1964), 192.

59. Henry Louis Gates, Jr., *The Signifying Monkey: A Theory of Afro-American Literary Criticism* (New York: Oxford Univ. Press, 1988), chap. 2. Krin Gabbard develops some of these concepts in "The Quoter and His Culture," in

Jazz in Mind: Essays on the History and Meanings of Jazz, ed. Reginald T. Buckner and Steven Weiland (Detroit: Wayne State Univ. Press, 1991), 92–111.

60. Interview with John Steiner, Milwaukee, Wisc., July 22, 1988.

61. Carmichael comments on Armstrong's tuition, in Hoagy Carmichael with Stephen Longstreet, *Sometimes I Wonder: The Story of Hoagy Carmichael* (New York: Farrar, Straus & Giroux, 1965), 203.

62. The Chicago *Defender,* April 16, 1927, p. 8, announces the publication of Armstrong's two books. Collier, *Louis Armstrong,* 178–79, discusses them.

Chapter Three

1. Isham Jones, "American Dance Music Is Not Jazz," *Etude* 42 (1924), 526.

2. A retrospective account on the rise of the dance craze in Chicago appeared in "Dance Craze Sweeps On—Loop Theaters Hurt," *Variety,* Jan. 5, 1923, p. 1. Its ramifications for the music business are discussed in New York *Clipper,* March 20, 1918, p. 12; March 5, 1919, p. 16; Dec. 17, 1919, p. 16; June 30, 1919, p. 17; June 21, 1924, p. 17.

3. Urban reformer Louise de Koven Bowen's *Our Most Popular Recreation Controlled by the Liquor Interests: A Study of Public Dance Halls* (Chicago: Juvenile Protective Association, 1911), 1–9; and Bowen, *The Public Dance Halls of Chicago* (Chicago: JPA, 1917), 4–9. Lewis A. Erenberg, *Steppin' Out: New York Nightlife and the Transformation of American Culture, 1890–1930* (Chicago: Univ. of Chicago Press, 1981), 151–58, dates New York's social dance craze from the 1910s.

4. Chicago *Herald,* Jan. 31, 1916, p. 16, compares Chicago cabarets with those in San Francisco; Chicago *Herald,* Jan. 24, 1916, p. 18, describes Frieberg's and Ike Bloom. See also Chicago *Herald,* Jan. 25, 1916, p. 16; Jan. 26, 1916, p. 3.

5. Louise de Koven Bowen, *The Straight Girl on the Crooked Path: A True Story* (Chicago: JPA, 1916), 10, 18–19; *The Road to Destruction Made Easy in Chicago* (Chicago: JPA, 1916), 11–13. The block quote is taken from Bowen, *The Public Dance Halls,* 5. Edward A. Berlin, *Reflections and Research on Ragtime* (Brooklyn: Institute for Studies in American Music, 1987), 3–5, describes ragtime saloons in New York as do Dempsey J. Travis, *An Autobiography of Black Jazz* (Chicago: Urban Research Institute, 1983) and Eileen Southern, *The Music of Black Americans* (New York: W.W. Norton, 1983) for Chicago, but the subject deserves a separate study. Jane Addams, *The Spirit of Youth and the City Streets* (New York: Macmillan, 1909) gives a general, critical appraisal. Joanna J. Meyerowitz, *Women Adrift: Independent Wage Earners in Chicago, 1880–1930* (Chicago: Univ. of Chicago Press, 1988) provides invaluable information on the kind of women who frequented dance halls in Chicago. Kathy Peiss, *Cheap Amusements: Working Women and Leisure in Turn-of-the-Century New York* (Philadelphia: Temple Univ. Press, 1986), 100–101, offers a comprehensive cul-

tural interpretation of women and leisure time activities in New York. The comments of Westbrook Pegler appear in Louis M. Starr, "45 Years of Night Life," Chicago *Sun,* June 6, 1943, p. 47.

6. Chicago's reformers admitted that the cabarets that sprang into business after 1916 were less vicious than those operating before Mayor Harrison's raids. See Herbert Asbury, *Gem of the Prairie: An Informal History of the Chicago Underworld* (New York: Knopf, 1940), 310. Louise de Koven Bowen summarizes high points of her life in *Growing Up with a City* (New York: Macmillan, 1926); also see Michael David Levin, "Louise de Koven Bowen: A Case History of the American Response to Jazz," Ph.D. diss. Univ. of Illinois at Urbana/Champaign, 1985, discusses Bowen's attitudes toward jazz. For a useful summary of the work of New York City dance hall reformers, see Elisabeth I. Perry, "'The General Motherhood of the Commonwealth': Dance Hall Reform in the Progressive Era," *American Quarterly* 37 (1985), 719–33.

7. The instructions for proper Victorian social dancing come from "Teresa Dolan Dancing Academy, 40th Street and Cottage Grove Avenue, Season 1914–15," in Dancing. Chicago. Miscellaneous Pamphlets, CHS. T. A. Faulkner, *From the Ball-Room to Hell* (Chicago: Henry Brothers, 1894), 1–10, 14, delivers a flamboyant diatribe detailing the ways in which social dancing led to drinking, sex, and character disorganization among young women. On popular dance styles, see also Bowen, *Safeguards for City Youth; The Road to Destruction; The Straight Girl on the Crooked Path;* and *The Public Dance Halls of Chicago.*

8. Reports on City Council efforts to regulate cabarets appear in the New York *Clipper,* April 3, 1918, p. 5; April 15, 1918, p. 14; June 12, 1918, p. 11; June 26, 1918, p. 11; Sept. 1, 1920, p. 5; April 6, 1921, p. 30; April 13, 1921, p. 7; April 20, 1921, p. 34. The ordinance ordering space utilization appears in *Report of the Proceedings of Chicago City Council, 1917–18* (Chicago: City of Chicago, 1918), 2510–11; extensive commentary on it by John Torman, chairman of the Committee on Licenses appears *ibid.,* 1476–80; New York *Clipper,* April 3, 1918, p. 5.

9. City of Chicago, *Report of the Proceedings of City Council, 1914–1915* (Chicago: City of Chicago, 1916), 2457–58, 3132; *Report of the Proceedings of City Council, 1915–16* (Chicago: City of Chicago, 1916), 871, 3234; *Report of the Proceedings of City Council, 1916–17* (Chicago: City of Chicago, 1917), 161, 641–42, 1011, 1892, 2079, 3587; *Report of the Proceedings of City Council, 1917–18,* 2510–11; 1476–80 contain the block quote, but also see 1810–12. *Report of the Proceedings of City Council, 1918–19* (Chicago: City of Chicago, 1919), 960–62; *Report of the Proceedings of City Council, 1919–1920* (Chicago: City of Chicago, 1920), 416, 1079, 1098, 2176; *Report of the Proceedings of City Council, 1921–22* (Chicago: City of Chicago, 1922), 1010–11, 1152. *Variety,* Dec. 29, 1922, p. 15.

10. James Lincoln Collier, *Benny Goodman and the Swing Era* (New York: Oxford Univ. Press, 1989), 31; George Bushell, Jr., "When Jazz Came to Chi-

cago," *Chicago History* 1 (1971), 132–41; Paul Eduard Miller, *Down Beat's Year-book of Swing* (Chicago: Downbeat Pub. Co., 1939), 7.

11. Miller, *Yearbook of Swing,* 7–8; Leonard Feather, "Jazz: Chicago Style," *Carte Blanche* (1961), 24–27, 52–53; Johnny Stein in *Hear Me Talkin' to Ya: The Story of Jazz as Told by the Men Who Made It,* ed. Nat Shapiro and Nat Hentoff (New York: Dover Pub., 1955), 82–84. Martin T. Williams, *Jazz Masters of New Orleans* (New York: DaCapo, 1979), 26–32. Chicago *Herald,* Jan. 24, 1916, p. 18. Collier, *Benny Goodman and the Swing Era,* 31. Leonard labels the earliest white jazz music as "nut" jazz in *Jazz and the White Americans: The Acceptance of a New Art Form* (Chicago: Univ. of Chicago Press, 1962), 13.

12. Brian Rust, *The American Dance Band Discography, 1917–1942,* I (New Rochelle, N.Y.: Arlington House, 1975), 169–73.

13. On Joe Frisco, see New York *Clipper,* March 20, 1918, p. 6; *ibid.,* Oct. 23, 1918, p. 9; Nov. 26, 1919, p. 11. Don DeMichael and Wayne Jones Taped Interview with Bud Freeman, Nov. 3, 1980, CJA. *Variety,* Feb. 6, 1920, p. 9. *Talking Machine World,* Sept. 15, 1926, p. 154, insists that dance steps were popularized through "the professional stage" and "the large dance floors."

14. Gilda Gray is described in Erenberg, *Steppin' Out,* 250–51. Fritzel, the Arsonia Cafe, Friars Inn, and Crawford and Rogers are described in Louis M. Starr, "45 Years of Night Life," Chicago *Sun,* June 6, 1943, p. 47; also see Fritzel obituary, Chicago *Daily Tribune,* Sept. 29, 1956, pt. 3, p. 12. Joan Crawford's movie dancing is mentioned in Arnold Shaw, *The Jazz Age: Popular Music in the 1920s* (New York: Oxford Univ. Press, 1987), 8–9. Bee Palmer's movements are traced in New York *Clipper,* Aug. 9, 1919, p. 6; May 18, 1921, p. 12; *Eddie Condon's Scrapbook of Jazz,* ed. Eddie Condon and Hank O'Neal (New York: St. Martin's Press, 1973), n.p., reproduces a Palmer publicity photo.

15. "Chicago by Night," *Variety,* Oct. 6, 1926, p. 32.

16. *Variety,* Dec. 16, 1925, p. 46. *Orchestra World* 1 (Dec. 1925), 16. Paul G. Cressy *The Taxi-Dance Hall: A Sociological Study in Commercialized Recreation and City Life* (Chicago: Univ. of Chicago Press, 1932), 192–93.

17. JPA and NABRMP policies on dance tempi are described in *Variety,* Sept. 1, 1926, p. 46, making references to earlier events of 1921.

18. Brian Rust, *The American Dance Band Discography, 1917–1942,* I (New Rochelle, N.Y.: Arlington House, 1975), 137–42; O'Sickey Col. Collier's comments, applied to the dance bands of B. A. Rolfe and Louis Panico, appear in *Louis Armstrong: An American Genius* (New York: Oxford Univ. Press, 1983), 239.

19. The relations between the JPA and the ballroom proprietors and managers are detailed in "By-Laws of the National Association of Ball Room Proprietors and Managers," carbon ts.; Binford to Harmon, carbon ts., Oct. 27, 1924; Binford to Eitel, car. ts., Oct. 28, 1924; Binford to [Edward] Diedrich [Midway

Gardens], car. ts., Oct. 27, 1924; Binfield to Miss Kendall and Mr. Lund [Merry Gardens], Messrs. Ashton and Byfield [White City], Messrs. Karzas and Sheey [Trianon Ballroom], car. ts., Oct. 29, 1924; Binfield to Eitel, Green, Johnson, McCormick, Plain, McGuire, Smilzdorf, all car. ts., all dated Oct. 30, 1924; Binford to Harmon, car. ts., Oct. 12, 1928 and June 21, 1929; Binford to Byfield and Donlevy, Oct. 27, 1928. JPA File #102: "National Association of Ball Room Proprietors & Managers—by laws, correspondence, minutes." JPA questionnaires and selected answers on dance hall policies are contained in JPA File #104: "Public Dance Halls—reports December 1917–October 1928." JPAR.

20. Information on Harmon's Dreamland is scattered through the following sources: E[lizabeth] L. Crandall Report on Dreamland, 11–10–[28?] JPA File No. 104 "Public Dance Halls—reports December 1917–October 1928," JPAR. The ethnic identity of Dreamland dancers is characterized in Report on New Majestic [Dance Hall], Jan. 4, 1924, JPA File No. 103 "Closed Dance Halls," JPAR. *Variety,* July 27, 1923, p. 23; Jan. 13, 1926, p. 46. New York *Clipper,* Feb. 16, 1921, p. 23. Kernfeld, ed., *Jazz Grove,* II, 198. On the physical setting of Harmon's Dreamland, see Lawrence Gushee to John Steiner, ts., n.d., Joints of Chicago File, JSC, and Scrapbooks of Louise de Koven Bowen, I, Manuscripts Division, CHS.

21. John Lax Interview with Willie Randall, Dec. 28, 1971, ts., COHC.

22. Interview with Charles Elgar, Elgar File HJATU. Elgar is mentioned as providing the music for a variety of refined South Side cultural events, including Binga's yearly parties, in Chicago *Broad Ax,* May 18, 1912, p. 1; Dec. 31, 1921, p. 1; Jan. 1, 1921, p. 1. *Variety,* Dec. 16, 1925, p. 46. On Keppard's drinking habits, see Onah Spencer, "Freddie Keppard," in "Negro Music and Musicians," HC-WBCPL. The Chicago *Defender,* Oct. 10, 1925, p. 7, documents Oliver's appearance at Harmon's Dreamland.

23. *Variety,* Dec. 16, 1925, p. 46.

24. The history of White City Amusement Park is quickly summarized by Perry R. Duis and Glen E. Holt, "Bright Lights, Hard Times of White City," *Chicago History* 27 (1978), 176–79. Fascinating announcements and publicity from White City during the twenties can be found in White City Scrapbook of Announcements ... etc, 1914–1933," CHS. Insights into the socioeconomic characteristics of White City crowds will be found in Chicago *Herald,* Feb. 7, 1916, p. 18. "White City Dance Hall—Public Dance Halls, Reports December 1917–October 1928," Juvenile Protective Association File No. 104, JPAR. Chicago *Herald,* Feb. 7, 1916, p. 18. "White City Twin Ballrooms, Chicago's Fun Center, n.d., Scrapbooks of Amusements, 1914–1933, CHS. *Variety,* Oct. 7, 1925, p. 50.

25. The history of the original Midway Gardens is summarized in the Midways Gardens Co., *Midway Gardens* [Chicago: R. F. Welsh, 1914] and Col-

lege Humanities Staff, *The Midway Gardens, 1914–1929* [Chicago: Humanities Staff, 1961], CHS. On the Midway Dancing Gardens, see *Midway Dancing Gardens Toddle News* I, 1 (June 15, 1923), xerox, Joints of Chicago File, Steiner Collection. Paul Kruty, "Pleasure Garden on the Midway," *Chicago History* 16 (1988), 4–27.

26. Descriptions of dancers appear in New York *Clipper,* Oct. 5, 1923, p. 25; *Variety,* Oct. 15, 1924, p. 38; and Feb. 17, 1926, p. 47. Elmer Schoebel to Leonard Feather, Schoebel Vertical File, IJS. Jazz musicians are located at Midway Dancing Gardens in John Steiner Interview with Floyd Town, ms., n.d., Towne File, JSC. Derek Coller, "Notes on Jess Stacy, 19 June 1969, ts., Stacy File, JSC. Barry Kernfeld, ed., *New Grove Dictionary of Jazz* (London: Macmillan, 1988), 201, hereafter cited as *Jazz Grove;* Steiner m.s. notes, Spanier File; Alma Hubner, "Muggsy Spanier," in *Selections from the Gutter: Jazz Portraits from "The Jazz Record,"* ed. Art Hodes and Chadwick Hansen (Berkeley: Univ. of California Press, 1977), 178–79; *Variety,* Feb. 17, 1926, p. 47. The Midway Gardens Orchestra sides have been reissued on Fountain/Retrieval, "The Midway Special," Fountain DFJ—115. Spanier, Stacy, Floyd O'Brien, and Teschemacher are placed at the Midway Gardens in Charles Edward Smith, "The Austin High School Gang," in Smith and Ramsey, *Jazzmen* (New York: Harvest Books, 1939), 170–71. Stacy's work at Midway Gardens is described in Leonard Feather, "A Band Pianist Who Came In from the Hot," Los Angeles *Times* Sunday magazine, May 18, 1975, p. 32; and Marty Grosz, "Frank Teschemacher: Biography and Notes on the Music," *Giants of Jazz: Frank Teschemacher* (Alexandria, Va.: Time-Life Records, 1982).

27. Columbia Records, Chicago, May 30, 1923, reissued on *The Midway Special,* Fountain-Retrieval Records DFJ-115.

28. On the construction, location, and programs of the Trianon, see *Variety,* June 30, 1922, p. 1; Dec. 1, 1922, pp. 1, 4; Dec. 15, 1922, p. 9; April 26, 1923, p. 9; Nov. 15, 1923, p. 4. On Whiteman at the Trianon, see New York *Clipper,* Nov. 22, 1922, p. 28; James T. Maher, liner notes to "The Great Isham Jones and His Orchestra," RCA Victor LPV-504, xerox, Isham Jones File, Steiner Collection. Nancy Banks, "The World's Most Beautiful Ballrooms," *Chicago History* 2 (1973), 206–15, provides a useful description of the Trianon and Aragon ballrooms.

29. On Whiteman's symphonic syncopation, see *Variety,* Dec. 13, 1923, p. 1; Jan. 3, 1924, p. 4. On his Aeolian Hall Concert, see the issue for Dec. 3, 1924, p. 32, and for Green's assessments, Nov. 26, 1924, p. 50, and Sept. 24, 1924, p. 26c. The New York *Clipper,* Aug. 10, 1923, p. 21 identifies the popularity of a "steady, hammering rhythm."

30. Rust, *American Dance Band Discography,* I, 1009–10; O'Sickey Col.

31. *Variety,* July 30, 1924, p. 38; Nov. 4, 1924, p. 35. Rust, *American Dance Band Discography,* II, 1109–12. O'Sickey Col.

32. Real Estate Publicity Folder, Mann's Million Dollar Rainbo Gardens, 1960, CHS; "Marigold Gardens, Misc., Chicago," CHS; *This Week in Chicago* (Oct. 31–Nov. 6, 1920), n.p., CHS; *Variety,* March 24, 1922, p. 12; Dec. 12, 1922, p. 31; Nov. 25, 1925, p. 44; Sept. 1, 1926, p. 46. Rust, *American Dance Band Discography,* II, 1913–15, 1960.

33. The interior design and the entertainment at Terrace Gardens are graphically represented in advertisement found in theater programs. See Cort Theater Program, July 1, 1922, p. 8, and Garrick Theater Program Aug. 24, 1924, n.p., Theater Programs, 1919–27, CJA. For a description of the admiring groups of street people gathered at the entrance, see Irle Waller, *Chicago Uncensored: Firsthand Stories about the Al Capone Era* (New York: Exposition Press, 1965), 56–57.

34. Harold Leonard, "The Hotel Dance Orchestra," *Orchestra World* 1 (Jan. 31, 1926), 5–6.

35. Harold Leonard, "The Hotel Dance Orchestra," *ibid.* (Feb. 28, 1926), 5–6.

36. On Isham Jones, see New York *Clipper,* Feb. 2, 1921, p. 23; Feb. 9, 1921, p. 23; May 17, 1922, p. 28; Oct. 12, 1923, p. 25; June 14, 1924, pp. 18, 22; and *Variety,* Sept. 22, 1922, p. 26; Dec. 31, 1924, p. 26B; March 18, 1925, p. 45; April 22, 1925, p. 38; May 27, 1925, p. 45; Chicago *Defender,* July 15, 1922. Descriptions of the College Inn appear in *Variety,* March 31, 1926, p. 47, and April 28, 1926, p. 31.

37. "Aunt Hagar's Children's Blues" (Br2358), "Henpecked Blues" (Br2479), "It's the Blues" (Br3027), "Land O' Lingo Blues" (Br2738), "Forgetful Blues" (Br 2531), Isham Jones Orchestra, Brunswick Records. I would like to thank collector Joel O'Sickey for playing, discussing, and taping these original 78 rpm records for me.

38. On Edgar Benson's control of Chicago's major hotel and supper club orchestras, see *Variety,* April 8, 1925, p. 43; April 22, 1925, pp. 34, 37; May 6, 1925, p. 49; July 8, 1925, p. 41. New York *Clipper,* May 15, 1924, p. 16.

39. *Variety,* Dec. 29, 1922, p. 14; April 19, 1923, p. 11; Aug. 27, 1924, p. 38; New York *Clipper,* Aug. 17, 1923, p. 21; Dec. 7, 1923, p. 22; Jan. 18, 1924, p. 23; May 9, 1923, p. 18; June 21, 1924, p. 23.

40. *Variety,* March 18, 1925, p. 45; April 29, 1925, p. 39; May 6, 1925, pp. 49, 50; May 27, 1925, p. 45; June 17, 1925, p. 34.

41. New York *Clipper,* Sept. 21, 1921, p. 18; Dec. 28, 1921, p. 18; March 15, 1922, p. 18; Aug. 16, 1922, p. 13. *Variety,* Jan. 5, 1923, pp. 9, 35; March 22, 1923, p. 5; Aug. 6, 1924, p. 37; Aug. 27, 1924, p. 38; Sept. 10, 1924, p. 39; Oct. 1, 1924, p. 28; Dec. 24, 1924, p. 36. *Talking Machine World,* Aug. 15, 1926, pp. 42, 72; Oct. 15, 1926, p. 114.

42. New York *Clipper,* Nov. 22, 1922, p. 28; May 2, 1923, p. 28; May 9, 1923,

p. 28; Aug. 6, 1924, p. 56. *Variety,* April 30, 1924, p. 40; May 14, 1924, p. 48; May 28, 1924, p. 7: July 2, 1924, p. 48.

43. *Variety,* April 22, 1925, p. 34; April 29, 1925, p. 39; June 10, 1925, p. 39; March 24, 1926, p. 41; June 30, 1926, p. 41.

44. Onah Spencer, "Freddie Keppard," in "Negro Music and Musicians," VHCWB, CPL, described Keppard's appearances with the black dance bands.

Chapter Four

1. "John W. Wickliffe's Ginger Orchestra Styled America's Greatest Jaz Combination," Indianapolis *Freeman,* Oct. 28, 1916, p. 5. Nat Shapiro and Nat Hentoff, eds., *Hear Me Talkin' to Ya: The Story of Jazz as Told by the Men Who Made It* (New York: Dover, 1955), chap. 8, employs the metaphor of the "Second Line" to describe the white Chicago jazzmen who, in their admiration of the music of the older, black pioneers, formed a "Second Line," not unlike the New Orleans youngsters who danced along behind their favorite marching bands.

2. Paul Eduard Miller, "Thirty Years of Chicago Jazz, Chapter One," *Esquire's 1946 Jazz Book,* ed. P. E. Miller (New York: Smith & Durrell, 1946), 7.

3. Jane Addams, *The Spirit of Youth and the City Streets* (New York: Macmillan, 1909), 3–13, 18–19, 69. Addams describes the roots of jazz age culture in Chicago with remarkable equanimity and insight. Irle Waller, *Chicago Uncensored: Firsthand Stories About the Al Capone Era* (New York: Exposition Press, 1965), 47–48, describes the ready availability of marijuana, opium, and other narcotics in turn-of-the-century Chicago.

4. *Annual Report of the Juvenile Protective Association* (Chicago: JPA, 1924), 15–26; *Annual Report of the Juvenile Protective Association* (Chicago: JPA, 1927), 19–25.

5. *Studs Lonigan* (New York: Vanguard Press, 1932).

6. Mezz Mezzrow and Bernard Wolfe, *Really the Blues* (Garden City, N.Y.: Anchor Books, 1972), 3–5. Don DeMichael, "Percussion's Dean Steps to a Lively Beat," Chicago *Tribune,* Nov. 12, 1978, pp. 23, 26.

7. Ralph Berton, *Remembering Bix: A Memoir of the Jazz Age* (New York: Harper & Row, 1974), 184–87.

8. Max Kaminsky with V. E. Hughes, *My Life in Jazz* (New York: Harper & Row, 1963), chap. 3; Eddie Condon with narration by Thomas Sugrue, *We Called It Music: A Generation of Jazz* (New York: Holt, 1947), 90–110.

9. Neil Leonard, *Jazz, Myth, and Religion* (New York: Oxford Univ. Press, 1987), has analyzed in detail the religious dimensions of jazz sensibilities and behavior patterns.

10. Richard Hadlock, *Jazz Masters of the Twenties* (New York: Collier Books, 1965), 106–7. On Beiderbecke, see Richard M. Sudhalter and Philip R.

Evans with William Dean-Myatt, *Bix: Man and Legend* (New Rochelle, N.Y.: Arlington House, 1974) and Berton, *Remembering Bix.* John Paul Perhonis, "The Bix Beiderbecke Story: The Jazz Musician in Legend, Fiction, and Fact," Ph.D. diss., Univ. of Minnesota, 1978. John T. Schenck, "Biographical Sketch of Bud Jacobson," *Jazz Session* (Nov. 1945), 3–8; John Steiner, "Jake's Chicago Travelogue: A Factual History of Bud Jacobson's Activities," *Storyville* (June 1966), 5–7.

11. Dominic Pacyga and Ellen Skerrett, *Chicago: City of Neighborhoods* (Chicago: Loyola Univ. Press, 1986), 199. William H. Miller, "Floyd O'Brien," *Three Brass* (Melbourne, Australia: W. H. Miller, 1945).

12. Benny Goodman and Irving Kolodin, *Kingdom of Swing* (New York: Frederick Ungar, 1961), chap. 1; James Lincoln Collier, *Benny Goodman and the Swing Era* (New York: Oxford Univ. Press, 1989), chap. 1.

13. Interview with Art Hodes in Dempsey J. Travis, *Autobiography of Black Jazz* (Chicago: Urban Research Institute, 1983), 381–87; Art Hodes and Chadwick Hansen, eds., *Selections from the Gutter: Portraits from the "Jazz Record"* (Berkeley: Univ. of California Press, 1977), 9–11.

14. Amy Lee, "Muggsy Knew He Was Hooked," *Down Beat,* April 15, 1943; Bill Russell interview with Spanier, ms., April 21, 1957, WRHJA; Alma Hubner, "Muggsy Spanier," in Hodes and Hansen, eds., *Selections from the Gutter,* 178–79.

15. Rudi Blesh, *Combo USA: Eight Lives in Jazz* (Philadelphia: Chilton, 1971), 134–60; Bruce Crowther, *Gene Krupa: His Life & Times* (Tunbridge Wells, Eng.: Spellmount Ltd., 1987), 32; Mezzrow and Wolfe, *Really the Blues,* 123–24. John T. Schenck, "Life History of Voltaire de Faut," *The Jazz Session* July–Aug. 1945, pp. 2–3, 13. Art Hodes, "A Talk with Volly DeFaut," *Down Beat,* Dec. 1, 1966, pp. 22–23.

16. Ralph Berton to Leonard Feather, Dec. 23, 1954, ts., Vic Berton Vertical File, IJS; Berton, *Remembering Bix.*

17. *Really the Blues,* 88–89.

18. Charles Edward Smith, "The Austin High Gang," in *Jazzmen,* ed. Charles Edward Smith and Frederic Ramsey, Jr. (New York: Harvest Books, 1939), 161–69; Hadlock, *Jazz Masters of the Twenties,* 106–13.

19. Pacyga and Skerrett, *Chicago,* 169–71; Mezzrow and Wolfe, *Really the Blues,* 89; Berton W. Peretti, "White Hot Jazz," *Chicago History* 17 (1988–89), 26–41.

20. Austin High School *Maroon and White,* 1920–28, SCCPL. Chip Deffaa, "Jimmy McPartland's Story, Part I," *Mississippi Rag* 14 (July 1987), 1; McPartland as quoted in Shapiro and Hentoff, *Hear Me Talkin' to Ya,* 118. Wayne Jones and Don DeMichael Interview with Bud Freeman, Nov. 3, 1980, untranscribed audio tape, CJA. This interview should be carefully compared with Bud Freeman as told to Robert Wolf, *Crazeology: The Autobiography of a Chicago Jazzman*

(Urbana: Univ. of Illinois Press, 1989), 2, where Freeman says that he was born in Austin.

21. Goodman, *Kingdom of Swing,* 31; Interview with Jimmy McPartland, IJS.

22. "Austin High School Gang," 162; Freeman, *Crazeology,* 3; Jones & DeMichael Interview with Freeman. Vladimir Simosko, "Frank Teschemacher: A Reappraisal," *Journal of Jazz Studies* 3 (1975), 28–53. Whitney Balliett, "Jazz: Little Davy Tough," *The New Yorker,* Nov. 18, 1985, pp. 160–62, 165–66. Tough's dissatisfaction with the musician's lifestyle is described in George T. Simon, "the life and death of davey tough," *Metronome,* Feb. 1949, p. 17. See also Simon, "Dave Tough," *Metronome* 6 (June 1937), 21. Leonard Feather, "The Dave Tough Story," *Down Beat,* July 1, 1953, describes Tough's "periodic wild masochistic jags," "his personality split between the desires to play and to write."

23. Kaminsky, *My Life,* 37–38.

24. Wayne Jones Interview with Wild Bill Davison, Sept. 10, 1981, CJA; author's Interview with Davison, Washington, D.C., April 14, 1976. Marty Grosz, "Frank Teschemacher: Biography and Notes on the Music," *Giants of Jazz: Frank Teschemacher* (Alexandria, Va.: Time-Life Records, 1982), 9; Freeman, *Crazeology,* 10.

25. Condon, *We Called It Music,* 102–4; William H. Kenney, III, "Eddie Condon in Illinois: The Roots of a Jazz Personality," *Illinois Historical Journal* 77 (1984), 255–68.

26. Hodes Interview in Travis, *Autobiography of Jazz,* 383; Kaminsky, *My Life,* 10–39. Condon, *We Called It Music,* 49; Jones/DeMichael Interview with Freeman, CJS.

27. Berton, *Remembering Bix,* 80, 120–21.

28. As quoted in Shapiro and Hentoff, *Hear Me Talkin' to Ya,* 120.

29. Elmer Schoebel to Leonard Feather, n.d., Vertical File, IJS; [George Hoefer] to Elmer Schoebel, June 19, 1942, ts., Vertical File, IJS.

30. Sudhalter and Evans, *Bix,* 63–68, 70–78; Condon, *We Called It Music,* 71–72.

31. Sudhalter and Evans, *Bix,* 135–39. *Variety,* Nov. 18, 1925, p. 47, describes George Leiderman and Sam Rothschild's Rendez-Vous cabaret; and *Orchestra World* 1 (June 1925), 12, locates it "in the Gold Coast section."

32. *Orchestra World* (Summer 1927), 8. *Variety,* Sept. 10, 1924, p. 40, describes the Wolverines' appearance at New York City's Cinderella Ballroom; the Sept. 24, 1924 issue, p. 26c, reports that they were "a torrid unit" and a "hit."

33. George Wettling, "A Tribute to Baby Dodds from George Wettling," *Down Beat* 29, 7 (March 29, 1962), 21; Richard Gehman, "George, the Legendary Wettling," *Jazz* 4 (Oct. 1965), 16–19; A. V. Baillie, "Kewpie Doll—A Warm Drummer," *Coda* 2 (Feb. 1960), 21–24. Hubner, "Muggsy Spanier," 178; Russell Interview with Spanier, WRHJA; Amy Lee, "Muggsy Knew."

34. Freeman, *Crazeology*, 6, 8, 10; Spencer in "Negro Music and Musicians," HCWB. Condon, *We Called It Music,* 92.

35. Condon, *ibid.,* 94; Spencer, "Negro Music;" Jones/DeMichael Interview with Freeman, CJA.

36. Travis, *Autobiography of Black Jazz,* 66, 69–71; Jones/DeMichael Interview with Freeman, CJA.

37. Souchon, "King Oliver: A Very Personal Memoir," in *Jazz Panorama: From the Pages of the Jazz Review,* ed. Martin T. Williams (New York: Crowell-Collier, 1962), 21–30. Condon, *We Called It Music,* 94; Travis, *Autobiography of Black Jazz,* 43, mentions Jim Crow seating patterns in the South Side black-and-tans. My interpretation of the descriptions later given by white jazzmen of their experiences in entering the South Side clubs owes much to Neil Leonard, *Jazz, Myth, and Religion,* chap. 4, and Victor Turner, "Liminal to Liminoid, in Play, Flow, and Ritual: An Essay in Comparative Symbology," *From Ritual to Theatre: The Human Seriousness of Play* (New York: PAJ Publications, 1982), 20–60, but particularly 24–28.

38. Mezzrow, *Really the Blues,* 23; Condon, *We Called It Music,* 91.

39. Chubb-Steinberg Orchestra, "Horsey, Keep Your Tail Up," April 10, 1924, OK 40107, should be compared with any of the sides Davison cut later.

40. Kaminsky, *My Life,* 39–41.

41. Balliett, "Jazz: Little Davy Tough"; Wettling, "A Tribute to Baby Dodds."

42. Collier, *Benny Goodman and the Swing Era,* 19, 61–64; Goodman and Kolodin, *Kingdom of Swing,* 26; Benny Goodman in Stanley Dance, *The World of Swing* (New York: Scribner's, 1974), 262. William Howland Kenney, III, "Jimmie Noone: Chicago's Classic Jazz Clarinetist," *American Music* 4 (1986), 145–58. Marsala is quoted in Shapiro and Hentoff, *Hear Me Talkin' to Ya,* 126–27.

43. Kenney, "Jimmie Noone," 154, 156.

44. Onah Spencer, "Negro Music and Musicians," HCWB.

45. Freeman, *Crazeology,* 7.

46. Travis, *Autobiography of Black Jazz,* 57–58; Jones Interview with Davison, CJA; Jacobson in Banks, *First-Person America,* ed. Ann Banks (New York: Knopf, 1980), n.p.

47. Piazza Interview with Hinton, IJS; Jones Interview with Davison, CJA.

48. John Lax Interview with Scoville Brown, Dec. 14, 1971, ts., COHC.

49. Bud Jacobson in *First-Person America;* Travis, *Autobiography of Black Jazz,* 66; Freeman, *Crazeology,* 6. Kaminsky, *My Life,* 40. Stanley Dance, *World of Earl Hines* (New York: Scribner's, 1977), 48. Richard B. Hadlock, "Joe Sullivan: Biography and Notes on the Music," *Giants of Jazz* (Alexandria, VA.: Time-Life Records, 1982).

50. Onah Spencer, "Negro Music and Musicians," HCWB. Peyton, "The Musical Bunch," Chicago *Defender,* Sept. 29, 1928, p. 10.

51. DeMichael Interview with Freeman; Art Hodes, "Liberty Inn Drag," in *Notes from the Gutter,* 23; Hoagy Carmichael, *Sometimes I Wonder* (New York: Farrar, Straus & Giroux, 1965), 102–3.

52. *Variety,* March 4, 1925, p. 45; Nov. 11, 1925, p. 48; Sept. 8, 1926, p. 46; Sept. 29, 1926, p. 53; Dec. 29, 1926, p. 37. "Merritt Brunies and His Friars Inn Orchestra, 1924–1926," Fountain Retrieval Records FJ-26.

53. "Art Hodes," in Travis, *Autobiography of Black Jazz,* 383; Art Hodes, "Everybody's in the Union," *Selections from the Gutter,* 11–12. Condon, *We Called It Music,* 96–100, 107–10. John T. Schenck, "Biographical Sketch of Bud Jacobson," *Jam Session* (Nov.–Dec. 1945), 4.

54. Lax Interview with Scoville Brown.

55. Dave Peyton, "The Musical Bunch," Chicago *Defender,* Dec. 12, 1925, p. 7.

56. Lil Hardin Armstrong as quoted in Jones/DeMichael Interview with Freeman, CJA.

57. Alan Lomax, *Mr. Jelly Roll: The Fortunes of Jelly Roll Morton, New Orleans Creole and "Inventor of Jazz"* (New York: Duell, Sloan, and Pearce, 1950), 181–82.

58. John Lax Interview with Willie Randall, Dec. 28, 1971, ts., COHC; Lax Interview with Red Saunders, Dec. 24, 1971, ts., COHC; Lax Interview with Ralph E. Brown, Dec. 29, 1971, ts., COHC; Lax Interview with Scoville Brown, COHC.

59. *Really the Blues,* 138–42. For an insightful discussion of primitivism and jazz, see: Ted Gioia, "Jazz and the Primitivist Myth," *Music Quarterly* 73 (1989), 130–43.

60. Norman Mailer, "The White Negro: Superficial Reflections on the Hipster," *Dissent* (1957), 276–93.

61. Bernard Wolfe, "The Ecstatic in Blackface," appended to *Really the Blues* (New York: Anchor Books, 1972), 293–304.

62. DeMichael and Jones Interview with Davison.

63. *Really the Blues,* 127–33; Condon, *We Called It Music,* 127–37; Freeman, *Crazeology,* 25–27.

64. Mezzrow, *Really the Blues,* 127–31. Robert S. Gold, *A Jazz Lexicon* (New York: Knopf, 1964), 161.

65. Freeman, *Crazeology,* 34.

Chapter Five

1. *Variety,* Nov. 5, 1924, 35.

2. Gunther Schuller, *Early Jazz: Its Roots and Musical Development* (New York: Oxford Univ. Press, 1968); Martin T. Williams, *The Jazz Tradition* (New York: Oxford Univ. Press, 1970), chaps. 2–4; Rudi Blesh, *Shining Trumpets: A*

History of Jazz, 2nd ed. (New York: Da Capo Press, 1976), chaps. 10–11; Hugues Panassié, *Hot Jazz* (New York: M. Witmark, 1936).

3. Roland Gelatt, *The Fabulous Phonograph: From Tin Foil to High Fidelity* (Philadelphia: Lippincott, 1955), 172, 188–90. The negative attitude of phonograph retailers toward phonograph records finds documentation in "Quick Profits in Sales of Records," *Talking Machine World* 24 (July 1928), 6.

4. Chicago *Defender,* Oct. 21, 1916, p. 5, calls for a survey of South Side phonograph ownership. Ronald Clifford Foreman, "Jazz and Race Records, 1920–32: Their Origins and Their Significance for the Record Industry and Society," Ph.D. diss., Univ. of Illinois, 1968, pp. 35–36. Stephen Calt, "The Anatomy of a 'Race' Label—Part One," *78 Quarterly* 1 (1988), 19; Calt and Gayle Dean Wardlow, "The Buying and Selling of Paramounts," *78 Quarterly* 1 (1990), 21, identifies the Artophone machine.

5. New York *Clipper,* March 22, 1922, p. 28. Gelatt, *Fabulous Phonograph,* 191, 208, 210, 212.

6. Foreman, "Jazz and Race Records," 12–37. All references to specific contemporary recording sessions and record issues have been drawn from Brian Rust, comp., *Jazz Records, 1897–1942,* rev. ed., 2 vols. (Chigwell, Eng.: Storyville Publications, 1975), and Rust, *The American Dance Band Discography, 1917–1942,* 2 vols. (New Rochelle, N.Y.: Arlington House, 1975).

7. On the professional importance of phonograph records to recording artists, see *Variety,* Aug. 2, 1923, p. 5; Sept. 3, 1924, p. 37; New York *Clipper,* Aug. 24, 1923, p. 13; March 6, 1924, p. 13; the pay scale at recording sessions is specified in, Baby Dodds with Larry Gara, *The Baby Dodds Story* (Los Angeles: Contemporary Press, 1959), 72, and in Stephen Calt, "Paramount, Part 2: The Mayo Williams Era," *78 Quarterly* 1 (1989), 28.

8. See Rust, *Jazz Records,* I, 132. The elements of chronology in this history of jazz recording in Chicago are largely based on the ongoing work of Harold H. Hartel, who has been compiling a chronological listing of all jazz recording sessions, identifying them by, among other things, location. See: "the H³ chrono-matrix file," *Record Research,* nos. 175–226 (1980–86), *passim.* For the early Benson, Jones, and Westphal jazz records, see Hartel, "chrono-matrix file," no. 175–76 (Sept. 1980), 10; no. 177–78 (Nov. 1980), 4–5.

9. *Variety,* July 8, 1925, p. 41.

10. *Talking Machine World,* May 15, 1923, 150, and July 15, 1923, pp. 102, 137, report on efforts to promote records of black music in black neighborhoods. Brian Rust, *The American Record Label Book* (New York: DaCapo Press, 1984), 212–17. Stephen Calt, "Paramount, Part 2," 10–30. On the implications of the term "race records," see also James Lincoln Collier, *Louis Armstrong: An American Genius* (New York: Oxford Univ. Press, 1983), 96–97; and Kathy J. Ogren, *The Jazz Revolution: Twenties America and the Meaning of Jazz* (New York: Oxford Univ. Press, 1989), 91–93.

11. Rust, *American Record Label Book*, 212–14; Foreman, "Jazz and Race Records," 57–58; Robert M. W. Dixon and John Godrich, *Recording the Blues* (New York: Stein and Day, 1970), 9. *Talking Machine World*, May 15, 1923, p. 150; July 15, 1923, pp. 102, 104; Nov. 15, 1923, pp. 110, 112.

12. Paul Oliver, *Songsters and Saints: Vocal Traditions on Race Records* (Cambridge, Eng.: Cambridge University Press, 1984), 1–17.

13. *Variety*, Aug. 2, 1923, p. 5. "Inside on Negro Blues Wave Explained by Writer," New York *Clipper*, Sept. 28, 1923, p. 24. Record store sales of blues records in Chicago receive comment in Oscar Hunter, "Negro Music in Chicago," Aug. 23, 1940, HCWB. Calt and Wardlow, "The Buying and Selling of Paramounts," 7–24, details the sales process for one important company.

14. Hunter "Negro Music"; *Variety*, June 16, 1922, p. 22; March 18, 1925, p. 51; New York *Clipper*, Aug. 24, 1923, p. 13, and Sept. 14, 1923, p. 31.

15. Chicago *Defender* (1923–24), *passim*. Hunter, "Negro Music."

16. The reaction to Europe is noted by Foreman, "Jazz and Race Records," 18, 20 and fn.31; but also see Chicago *Defender*, May 10, 1919, p. 20.

17. Chicago *Defender*, July 31, 1920, p. 4.

18. *Ibid.*, May 5, 1923, p. 6. The commercial promise of the Okeh "Colored Folder" is reported in New York *Clipper*, March 15, 1922, p. 28.

19. *Defender*, Nov. 29, 1924, p. 6.

20. *Talking Machine World*, March 15, 1926, p. 104. *Defender* Jan. 30, 1926, p. 7; Feb. 27. 1926, p. 7.

21. *Talking Machine World*, May 15, 1926, p. 18. *Defender*, May 8, 1926, p. 6; May 15, 1926, p. 6; June 5, 1926, p. 7; June 12, 1926, p. 6; June 19, 1926, p. 6.

22. *Variety*, June 2, 1923, p. 5, and June 16, 1922, p. 22.

23. New York *Clipper*, Sept. 14, 1923, p. 31.

24. James Lincoln Collier, *The Reception of Jazz in America: A New View* (Brooklyn, N.Y.: Institute for Studies in American Music, 1988), 13, describes the Kapp store. On Brown's, see Chicago *Defender*, March 9, 1929, p. 10.

25. Chicago *Broad Ax*, Feb. 19, 1921, p. 1.

26. Calt and Wardlow, "Paramount, Part 2."

27. *Ibid.*

28. *Ibid.*

29. Danny Barker, *A Life in Jazz*, ed. Alyn Shipton (New York: Oxford Univ. Press, 1986), 157–58. The New York *Clipper*, March 13, 1924, p. 16, comments on Mamie Smith's drop in popularity.

30. The larger South Side theater and dance hall orchestras are discussed by Thomas Hennessey, "From Jazz Age to Swing: Black Musicians and Their Music, 1917–1935," Ph.D. diss., Northwestern Univ., 1973; and Thomas J. Hennessey, "The Black Chicago Establishment 1919–1930," *Journal of Jazz Studies* 2 (Dec. 1974), 15–45.

31. Dave Peyton, "Recording Units," Chicago *Defender*, March 9, 1929, p. 10. Peyton makes similar accusations of broader scope in *Defender*, May 21, 1927, p. 8; May 21, 1927; Oct. 13, 1928, p. 9.

32. Harry Pace describes his determination to preserve "our race music and ... the wonderful voices and musical talent we have in the race," in Chicago *Broad Ax*, Feb. 19, 1921, p. 1. The broad range of music Pace recorded is indicated *ibid.*, May 7, 1921, p. 1.

33. *Variety*, Jan. 5, 1927, p. 51.

34. George W. Kay, "Those Fabulous Gennetts," *The Record Changer* (June 1953), 4–12; Rust, *Record Label Book*, 129–32. I am indebted to record collector and percussionist Joel O'Sickey of Cuyahoga Falls, Ohio, who provided the taped copies of hundreds of rare Chicago Jazz records upon which this chapter is largely based. I alone am responsible, however, for the interpretations given to these sources.

35. Charles Edward Smith, "White New Orleans," in *Jazzmen,* ed. Charles Edward Smith and Frederic Ramsey, Jr. (New York: Harvest Books, 1939), 56–58; Martin T. Williams, "N.O.R.K.," *Jazz Masters of New Orleans* (New York: Da Capo, 1979), 121–35. "New Orleans Rhythm Kings," *Jazz Grove,* II, 169–71.

36. For a definition of "head arrangement," see *Jazz Grove* I, 32–33.

37. Although it is possible to hear the influence of "Jazzin' Babies Blues" (Oliver's Jazz Band, Okeh 4975, June 23, 1923) in "Tin Roof Blues" (NORK Gennett 5105, July 17, 1923) the latter shows major rhythmic alterations to the former's shorter, clipped theme. NORK recorded Morton's "Wolverine Blues" on March 13, 1923 (Gennett 5102) and "Marguerite," "Mr. Jelly Lord," and "London Blues" on July 17 and 18, 1923. On Richard M. Jones, see George Hoefer, Jr., "Richard M. Jones," *The Jazz Session* (Jan. 1946), 2–3, 14.

38. Baby Dodds's comments appear in *The Baby Dodds Story,* 69. Kay, "Those Fabulous Gennetts," 4–12.

39. *Talking Machine World,* May 15, 1923, p. 120; June 15, 1923, p. 124; Aug. 15, 1923, p. 103; "Chicago Becoming a Recording Center," Nov. 15, 1923, pp. 110, 112. Also see Chicago *Defender* 6–23–23, 7, and Foreman, "Jazz and Race Records," 152–53. The important Oliver records on Okeh are reissued on "King Oliver's Jazz Band, 1923," Smithsonian Records R001 with insightful notes by Lawrence Gushee.

40. "Gets in Touch with 30,000 Colored Elks," *Talking Machine World,* Sept. 15, 1923, p. 102. *Talking Machine Journal,* Nov. 1923, p. 84.

41. Condon and Thomas Sugrue, *We Called It Music: A Generation of Jazz* (New York: Holt, 1947), 91. Condon's insight did not carry through to his own music, which was to remain highly impromptu.

42. In chronological order, the three Oliver recordings of "Dippermouth Blues" occurred on April 6 and June 23, 1923, and (as "Sugar Foot Stomp") on May 29, 1926. The Oliver band recorded "Snake Rag" for Gennett on April 6,

1923, and again for Okeh on June 22, 1923. Rust, *Jazz Records*, II, 1229, 1231.

43. Strategic departures from Oliver's Jazz Band to his Dixie Syncopators can be heard on "King Oliver and His Dixie Syncopators 1926," vol. 1, Swaggie Records 821.

44. *The Baby Dodds Story*, 71, documents discussions between band leaders and recording executives prior to the recording session itself.

45. Morton's Red Hot Peppers as listed in Rust, *Jazz Records*, II, 1162–63, and as reproduced on "Jelly Roll Morton and his Red Hot Peppers (1926–1927), Volume 3," RCA Victor, French Black and White Series 731 059.

46. The Bucktown Five, Gennett Records, Richmond, Ind., Feb. 25, 1924. Rust, *Jazz Records*, I, 214.

47. Mezz Mezzrow discusses the head arrangements on recordings by McKenzie and Condon's Chicagoans in Mezz Mezzrow and Bernard Wolfe, *Really the Blues* (Garden City, N.Y.: Anchor Books, 1972), appendix one.

48. *Giants of Jazz: Frank Teschemacher* (Alexandria, VA.: Time-Life Records, 1982) includes the major recordings by the white rebel Chicagoans and the insightful booklet "Frank Teschemacher: Biography and Notes on the Music" by Marty Grosz.

49. James Collier writes that Louis Armstrong's Hot Five sessions in Chicago in 1925 "were put together in the most casual fashion ... Lil, or Louis, or somebody else in the band would scribble out a few tunes ...," *Louis Armstrong: An American Genius* (New York: Oxford Univ. Press, 1983), 170. "The Louis Armstrong Story," vols 1–3, Columbia Records CL851-853.

50. Reuben Reeves and His River Boys recorded fifteen numbers in Chicago for Vocalion from June 10 to Oct. 1, 1929. Rust, *Jazz Records*, II, 1349–50. "Reuben 'River' Reeves and His Tributaries/River Boys, 1929," Fountain Retrieval Records FJ-126.

51. Rust, *Jazz Records*, II, 1542–43; Jabbo Smith's Chicago records are reissued on "The Ace of Rhythm, Jabbo Smith," MCA-1347.

52. A perceptive discussion of Hines's style will be found in Richard Hadlock, *Jazz Masters of the Twenties* (New York: Collier Books, 1965), 50–75. Hines's solo recordings were made for the QRS company in Long Island City, N.Y., on Dec. 8, 9, 12, 1929. Rust, *Jazz Records*, I, 797–98.

53. Noone's Apex Club Orchestra reissued on "Jimmie Noone and Earl Hines at the Apex Club, 1928," vol. 1, French MCA 510.039; "Jimmie Noone: Chicago Rhythm, 1928–1930," vol. 2, MCA 510.110.

54. "Archive of Jazz: Bix Beiderbecke," BGY Records 529.054; Rust, *Jazz Records*, II, 1860–61; Schuller, *Early Jazz*, 186–94, attributes Beiderbecke's style to his personality.

55. Chicago, Okeh Rec., Dec. 5, 12, 1928. Rust, *Jazz Records*, I, 54–55; Schuller, *Early Jazz*, 127–31.

56. As quoted in Calt and Wardlow, "Paramount, Part 2," 16–17.

57. Rust, *Jazz Records,* II, 1274–76. Parham has recently been brought to the attention of jazz record collectors through "Tiny Parham and His Musicians, 1926–1930," 3 vols., Swaggie Records 831–833.

58. Erskine Tate's Vendome Orchestra, "Static Strut," Vocalion 1027, reissued in *Giants of Jazz: Louis Armstrong* (Alexandria, Va.: Time-Life Records, 1978).

59. Carroll Dickerson's Savoy Orchestra, Chicago, *c.* May 25, 1928. Rust, *Jazz Records,* I, 436; reissued on "Chicago in the Twenties," Arcadia Records 2011.

60. Cook's Dreamland Orchestra, Chicago, July 10, 1926, Col. 727-D and 813-D, reissued on "Chicago in the Twenties, 1926–1928," Arcadia 2011. Cookie's Gingersnaps, Chicago, June 22, 1926, Okeh 8390 and 8369. O'Sickey Col.

61. Gunther Schuller, *The Swing Era: The Development of Jazz, 1930–1945* (New York: Oxford Univ. Press, 1989), 263–75.

62. Earl Hines and His Orchestra, Chicago, Victor Records, Feb. 22, 1929, O'Sickey Col; Rust, *Jazz Records,* I, 798.

63. Chicago, Victor Rec., Dec. 17, 1926; Rust, *Jazz Records,* II, 1298.

Chapter Six

1. *Variety,* Oct. 30, 1929, p. 1; Nov. 6, 1929, p. 1.

2. "Chicago," *Variety,* Oct. 7, 1925, pp. 5, 50; Oct. 14, 1925, p. 48; Oct. 28, 1925, p. 40; Dec. 15, 1926, p. 2.

3. Chicago *Defender,* Oct. 11, 1924, p. 6.

4. *Variety,* Oct. 7, 1925, p. 50; Dec. 2, 1925, p. 47; April 21, 1926, p. 45.

5. *Variety,* Jan. 19, 1927, p. 47; March 2, 1927. The trade papers carried no further articles on the cases, which probably never came to trial.

6. Chicago *Defender,* April 17, 1926, p. 6; March 12, 1927, p. 10; March 26, 1927, p. 8; April 2, 1927, p. 8; April 16, 1927, p. 8; April 23, 1927, p. 8; May 21, 1927, p. 9; June 4, 1927, p. 10; June 25, 1927, p. 8; July 2, 1927, p. 8.

7. *People of the State of Illinois (ex rel. Irving Cohen) v. City of Chicago,* Cook County Superior Court transcript, Dec. 6, 1924, CHS. The Entertainers Cafe, ordered closed by a Cook County judge, managed to remain open for several months on a writ of *supersedeas* protecting its liberty to do business until appeals had been heard in the U.S. Circuit Court. *Variety,* April 7, 1922, p. 18; May 24, 1923, p. 10; Oct. 29, 1924, p. 35. New York *Clipper,* Feb. 8, 1924, p. 15.

8. Barney Bigard, *With Louis and the Duke: The Autobiography of a Jazz Clarinetist,* ed. Barry Martyn (New York: Oxford Univ. Press, 1988), 31; Chicago *Defender,* July 2, 1927, p. 10; Oct. 8, 1927, p. 8, names Al Capone as co-proprietor of the Cafe de Paris; April 21, 1928, p. 8. Bottoms' roadhouse is mentioned in *Defender,* Sept. 5, 1931, p. 7.

9. *Variety*, Sept. 29, 1926, p. 53; Dec. 15, 1926, p. 2; Jan. 19, 1927; May 4, 1927, p. 52.

10. "U.S. Wars on Chicago Night Life," Chicago *Herald and Examiner*, Feb. 6, 1928, p. 1. *Variety*, June 22, 1927, p. 58; Oct. 26, 1927, p. 57; Nov. 30, 1927, p. 56; Feb. 8, 1928, p. 55; Feb. 15, 1928, pp. 54, 62. The March 28, 1928, issue, p. 37, declares Chicago's greatest cabaret era over. Other interpretations of the origins of the word "hip" refer to sources which appeared after the "hip rulings." Robert S. Gold, *A Jazz Lexicon* (New York: Knopf, 1964), 145–46; Robert S. Gold, *Jazz Talk* (Indianapolis: Bobbs-Merrill, 1975), 128.

11. *Variety*, March 21, 1928, p. 72; April 4, 1928, p. 56.

12. *Variety*, April 11, 1928, p. 56; *Defender*, March 2, 1929, p. 8; Rob Roy, "Cafes, Like Your Sunday Suit, Abandoned on Weekdays," *Variety*, June 25, 1932, p. 8. Lewis A. Erenberg, "From New York to Middletown: Repeal and the Legitimization of Nightlife in the Great Depression," *American Quarterly* 38 (1986), 761–78, attributes these changes to the repeal of prohibition, but the serious enforcement of prohibition itself also stimulated larger, dry, ballroom-like cabarets. The "[racial] restrictions" at Al Quodbach's Granada Cafe are confirmed in John Lax Interview with Scoville Brown, Dec. 14, 1971, ts., COHC.

13. *Variety*, Sept. 26, 1928, p. 57; March 28, 1928, pp. 1, 37.

14. William Russell, "Boogie Woogie," in *Jazzmen*, ed. Frederic Ramsey, Jr., and Charles Edward Smith (New York: Harvest Books, 1939), chap. 8. Robert Sylvester, *A Left Hand Like God* (New York: Da Capo, 1989).

15. Chicago *Defender*, Dec. 29, 1928, p. 10; Feb. 23, 1929, p. 8; Aug. 31, 1929, p. 9. Stanley Dance, *The World of Earl Hines* (New York: Scribner's, 1977), 57, 70–72. Chicago *Defender*, June 11, 1932, p. 6, laments the lack of cabarets on the South Side as does "Cafes, Like Your Sunday Suit," p. 8.

16. Chicago *Defender* April 19, 1928, p. 11; Sept. 29, 1928, p. 10; Nov. 3, 1928, p. 11; Dec. 15, 1928, p. 10; Feb. 2, 1929, p. 11; March 2, 1929, p. 8; March 23, 1929, p. 10; June 15, 1929, p. 10; July 13, 1929, p. 9; Aug. 31, 1929, p. 9; April 4, 1931, p. 9; May 9, 1931, p. 8; May 16, 1931, p. 8; July 14, 1934. *Orchestra World* (April 1932), 9.

17. Baby Dodds, as told to Larry Gara, *The Baby Dodds Story* (Los Angeles: Contemporary Press, 1959), 51–68. Stanley Dance Interview with Marge and Zutty Singleton, Washington, D.C., May 1975, ts, IJS.

18. "Orchestras Beware!" Chicago *Defender* Dec. 31, 1927, p. 6; Feb. 19, 1927, p. 8; July 14, 1928, p. 8; July 21, 1928, p. 10; July 28, 1928, p. 8; Aug. 18, 1928, p. 11.

19. *Variety*, Oct. 12, 1927, pp. 54, 58.

20. Information on the battles of the [white] local #10 of A.F.M. with the vaudeville/movie theaters appears in *"Guarantee Trust Company of New York v. National Theaters Corporation,* in Re: Chicago Federation of Musicians, an Order of the District Court of the U.S., Northern District of Illinois, Eastern Division,

CHS; *Intermezzo,* Official Journal of the Chicago Federation of Musicians, Local 10 of A.F.M., May 1927–April 1928, CHS; Robert D. Leiter, *The Musicians and Petrillo* (New York: Bookman Associates, 1953), 45–46; and Paul Herman Apel, "A Study of the Chicago Federation of Musicians Local No. 10, A.F. of M.," unpub. M.A. thesis, Univ. of Chicago, 1951. The African-American side of the experience emerges from Chicago *Defender,* July 14, 1928, p. 9; July 28, 1928, p. 8; Sept. 1, 1928, p. 9; Sept. 8, 1928, p. 8; Jan. 19, 1929, p. 10.

21. Neil Leonard, *Jazz and the White Americans: The Acceptance of a New Art Form* (Chicago: Univ. of Chicago Press, 1962), 102–5. *Variety,* Feb. 11, 1925, p. 33; May 6, 1925, p. 48.

22. *Variety,* Aug. 7, 1929, p. 228; Edwin P. Scheuing, "The Made-to-Order Orchestra," *Variety,* Sept. 4, 1929, p. 73. Chicago *Defender,* Oct. 5, 1929, p. 10.

23. Walter Barnes, "Hittin' High Notes," Chicago *Defender,* Jan. 9, 1932, p. 7. Bruce A. Linton, "A History of Chicago Radio Programming, 1921–1931 with Emphasis on Stations WMAQ and WGN," unpub. Ph.D. diss., Northwestern Univ., 1953.

24. Chicago *Defender,* Oct. 31, 1925, p. 6; Dec. 12, 1929, p. 8; March 12, 1927, p. 8; May 21, 1927, p. 8.

25. *Ibid.,* Dec. 8, 1928, p. 10; Feb. 9, 1929, p. 8.

26. *Ibid.,* Aug. 6, 1927, p. 10; Jan. 7, 1928, 5; Jan. 21, 1928, p. 8; Jan 28, 1928, p. 8; Feb. 4, 1928, p. 8; Feb. 18, 1928, p. 8; March 3, 1928, p. 10.

27. "Cheap Road House Parties Ruined Chicago's Night Clubs," *Variety,* May 11, 1927, p. 1; May 16, 1928, p. 1; May 22, 1929, p. 65. The composite portrait appears in Juvenile Protective Association File No. 103 "Public Dance Halls—Closed, Dec. 1923–Feb. 1929," 2. A general discussion of the phenomena appears in JPA File No. 105—"Roadhouse Survey of Cook County, Illinois, July–August 1929, Special Collections, University Library, Univ. of Illinois at Chicago.

28. Daniel Russell, "The Road House: A Study of Commercialized Amusements in the Environs of Chicago," M.A. thesis, Univ. of Chicago, 1931, pp. 11, 118, 167–68. Jessie Binford, "May We Present the Roadhouse?," *Welfare Magazine* (July 1927), 5–6.

29. JPA File #105—Roadhouse Survey of Cook Co.

30. Leroy Ostransky, *Jazz City: The Impact of Our Cities on the Development of Jazz* (Englewood Cliffs, N.J.: Prentice-Hall, 1978), chap. 9.

31. James Lincoln Collier, *Benny Goodman and the Swing Era* (New York: Oxford Univ. Press, 1989), chap. 4.

32. Jimmy McPartland, as quoted in *Hear Me Talkin' to Ya: The Story of Jazz as Told by the Men Who Made It,* ed. Nat Shapiro and Nat Hentoff (New York: Dover Publications, 1955), 278–79. Bud Freeman as told to Robert Wolf, *Crazeology: The Autobiography of a Chicago Jazzman* (Urbana: Univ. of Illinois Press, 1989), 31–32.

33. Eddie Condon with narration by Thomas Sugrue, *We Called It Music: A Generation of Jazz* (New York: Holt, 1947). Among the important New York recording sessions which established the Chicago style were: Eddie Condon Quartet, July 28, 1928; Eddie Condon and His Footwarmers, Oct. 30, 1928; and Eddie's Hot Shots, Feb. 8, 1928; and Billy Banks and His Orchestra, April 18, May 23, July 26, 1932.

34. *Variety*, May 18, 1927, 1, p. 52.

35. Thomas J. Hennessey states that "disaster struck" for black Chicago bands in the dance halls when Charles Cook severed relations with Harmon's Dreamland in order to avoid being fired and Charles Elgar lost his long-standing position at Harmon's Arcadia Ballroom. "From Jazz Age to Swing: Black Musicians and Their Music, 1917–1935," Ph.D. diss., Northwestern Univ., 1973, pp. 339–40. Chicago *Defender* Jan. 19, 1929, p. 10; Jan. 26, 1929, p. 10; March 13, 1929, p. 10; April 9, 1927, p. 9; May 7, 1929, p. 8; May 11, 1929, p. 9; July 23, 1927, p. 6; Sept. 3. 1927, p. 8; Sept. 17, 1927, p. 7; Oct. 15, 1927, p. 9; Oct. 22, 1927, p. 9.

36. *Variety*, March 7, 1928, p. 57; Sept. 26, 1928, pp. 1, 58.

37. "Savoy Is Ballroom Beautiful," Chicago *Herald and Examiner*, Feb. 4, 1928, p. 9. *Variety*, April 5, 1923, p. 47; May 14, 1924, p. 48; July 2, 1924, p. 48; Oct. 1, 1924, p. 1; Feb. 4, 1925, p. 37; Feb. 11, 1925, p. 35. Chicago *Defender*, Nov. 12, 1927, p. 2; Jan. 21, 1928, pp. 8–9; Jan. 28, 1928, p. 8; April 28, 1928, p. 8.

38. Chicago *Defender*, Feb. 4, 1928, p. 9; Feb. 11, 1928, p. 9; March 23, 1929, p. 10; March 30, 1929, p. 9; April 19, 1928, pp. 10–11; May 5, 1928, p. 10.

39. Chicago *Defender*, April 5, 1928, p. 11; April 26, 1928, p. 10; Feb. 2, 1929, p. 10.

40. Chicago *Defender*, Jan. 28, 1928, p. 9; April 5, 1928, pp. 10–11; April 19, 1928, p. 11; May 11, 1929, p. 9.

41. Chicago *Defender*, June 30, 1928, p. 11; July 14, 1928, p. 9; Oct. 13, 1928, p. 10; Feb. 2, 1929, p. 11; March 23, 1929, p. 10; March 30, 1929, p. 9; April 21, 1929, p. 8. James Lincoln Collier, *Louis Armstrong: An American Genius* (New York: Oxford Univ. Press, 1983), 165.

42. Hines's activities both at the Grand Terrace and on the road are mentioned so often in the Chicago *Defender* that individual citations are impossible. See Stanley Dance, *The World of Earl Hines* (New York: Scribner's, 1977), 2, 48–63, 74–85.

43. Chicago *Defender*, Jan. 30, 1932, p. 7; Feb. 13, 1932, p. 6; Feb. 27, 1932, p. 6; March 5, 1932, p. 8.

44. Art Hodes, "Liberty Inn Drag," in *Selections from the Gutter: Portraits from the "Jazz Record,"* ed. Hodes and Chadwick Hansen (Berkeley: Univ. of California Press, 1977), 19–21. Hodes, "Jam Session," *ibid.*, 21–23.

45. Bud Freeman, *Crazeology*, chap. 4; Max Kaminsky with V. E. Hughes, *My Life in Jazz* (New York: Harper & Row, 1963), 62–66.

46. As quoted in Samuel B. Charters, *Jazz: A History of the New York Scene* (New York: Da Capo Press, 1981), 155–56.

47. Gilbert Millstein, "For Kicks, I" *The New Yorker* (March 9, 1946), 30–34, 36–37; "For Kicks, II," *ibid.* (March 16, 1946), 34–38, 41–43.

48. George Frazier, "Someday Nick's Might Be the Hallowed Home of Jazz," *Down Beat* 8 (Nov. 15, 1941), 6; Mary Peart, "Home of Dixieland Jazz," *The Jazz Session* 9 (1945), 3–10; "Nick Rongetti Died in New York on July 25," *Down Beat* 13 (Aug. 12, 1946), 2.

49. Author's interview with Milt Gabler, New Rochelle, N.Y., untranscribed audio tape, July 28, 1976. "Chicago Jazz Album: Featuring All Star Personnel," Decca Records DL 8029.

50. Condon, *We Called It Music*. Eddie Condon and Hank O'Neal, *The Eddie Condon Scrapbook of Jazz* (New York: St. Martin's Press, 1973).

51. Hugues Panassié, *Hot Jazz: The Guide to Swing Music,* trans. Lyle & Eleanor Dowling (New York: M. Witmark, 1934), 139–60. Panassié later revised his high estimate of the white Chicagoans in *The Real Jazz* (New York: Smith & Durrell, 1942), 63. Dave Dexter, Jr., *Jazz Cavalcade: The Inside Story of Jazz* (New York: DaCapo, 1977), chap. 4, derides the "Chicago Style" myth and attributes the idea to Hugues Panassié, Charles Delaunay, and George M. Avakian. Also see Robert G. White, "Chicago Style?" It's a Phony Myth!," *Music and Rhythm* (March 1941); 35–40.

52. *American Jazz Music* (New York: W. W. Norton, 1939), 125–27.

53. Rudi Blesh, *Shining Trumpets: A History of Jazz* (New York: Knopf, 1946), 237–38. Gunther Schuller, *Early Jazz: Its Roots and Musical Development* (New York: Oxford Univ. Press, 1968), 194n. Martin Williams, *The Jazz Tradition* (New York: Oxford Univ. Press, 1970) and Williams, ed. and comp. *The Smithsonian Collection of Classic Jazz* (Washington, D.C., Smithsonian Institution, 1973).

54. Victor Turner, *Dramas, Fields, and Metaphors: Symbolic Action in Human Society* (Ithaca: Cornell Univ. Press, 1974), 29–56, 272–86. Turner coins the term "liminoid" to describe similar emotional experiences in Western urban industrial societies, in *From Ritual to Theater: The Human Seriousness of Play* (New York: PAJ Publications, 1982), 20–60.

Index